Lean Six Sigma
for Hospitals

About the Author

Jay Arthur, The KnowWare Man, works with companies that want to boost profits using Lean Six Sigma. He is the author of *Lean Six Sigma Demystified* and the QI Macros Lean Six Sigma software for Excel, which is used by more than 3,000 hospitals. Jay has worked with Tenet Healthcare, Kindred Healthcare, Centura Healthcare, and Christus Healthcare on projects to improve patient flow and reduce clinical and operational errors.

Lean Six Sigma for Hospitals

SIMPLE STEPS TO FAST, AFFORDABLE, FLAWLESS HEALTHCARE

JAY ARTHUR

New York   Chicago   San Francisco   Lisbon   London
Madrid   Mexico City   Milan   New Delhi   San Juan
Seoul   Singapore   Sydney   Toronto

The *McGraw·Hill* Companies

Library of Congress Cataloging-in-Publication Data

Arthur, Jay, author.
  Lean six sigma for hospitals : simple steps to fast, affordable, flawless healthcare / Jay Arthur.
      p. ; cm.
  Includes bibliographical references and index.
  ISBN 978-0-07-175325-8 (pbk. : alk. paper)
  1. Hospitals—Administration. 2. Organizational effectiveness. 3. Medical care—Quality control. 4. Total quality management. I. Title.
  [DNLM: 1. Efficiency, Organizational. 2. Hospital Administration—methods. 3. Quality of Health Care—organization & administration. 4. Total Quality Management—methods. WX 150.1]
  RA971.A825 2011
  362.11068—dc22

                                                                              2010050751

McGraw-Hill books are available at special quantity discounts to use as premiums and sales promotions, or for use in corporate training programs. To contact a representative please e-mail us at bulksales@mcgraw-hill.com.

Lean Six Sigma for Hospitals

3 4 5 6 7 8 9 0   QFR/QFR   1 9 8 7 6 5 4 3 2

ISBN   978-0-07-175325-8
MHID   0-07-175325-7

The pages within this book were printed on acid-free paper.

Sponsoring Editor Judy Bass	**Proofreader** Paul Tyler
Acquisitions Coordinator Michael Mulcahy	**Indexer** Judy Davis
Editorial Supervisor David E. Fogarty	**Production Supervisor** Pamela A. Pelton
Project Manager Patricia Wallenburg	**Composition** TypeWriting
Copy Editor James Madru	**Art Director, Cover** Jeff Weeks

Contents

CHAPTER 3

Simple Steps to a Better Hospital 53

CHAPTER 4

Reducing Defects with Six Sigma 81

CHAPTER 5

Simple Steps to a Cheaper
(More Profitable) Hospital 117

CHAPTER 6

Six Sigma for Hospitals 135

CHAPTER 7

Excel Power Tools for Lean Six Sigma 155

CHAPTER 8

Is There an Improvement Project
in My Data? 189

CHAPTER 9

Sustaining Improvement 205

CHAPTER 10

Laser-Focused Process Innovation 221

CHAPTER 11

Statistical Tools for Lean Six Sigma 235

CHAPTER 12

Implementing Lean Six Sigma in Hospitals . .269

Preface

All organizations are sick.

—BOAZ RONEN,
Author, *Focused Operations Management for
Health Service Organizations*

*Healthcare is a terminal illness for America's governments
and businesses. We are in big trouble.*

—CLAYTON M. CHRISTENSEN,
Author, *The Innovator's Prescription:
A Disruptive Solution for Health Care*

Healthcare professionals are doing the best they can, but there is tremendous room for improvement. In 2005, H. James Harrington reported that an estimated 2.2+ million people around the world die as a result of a healthcare error every year. Every year, 2 million patients in the United States get an infection while hospitalized. Every six minutes, one of these patients dies as a result of a hospital-acquired infection (88,000 people per year), adding $5 billion to the cost of healthcare; 95 percent of these cases are preventable (83,500). 2010 estimates show that 99,000 people die of a hospital-acquired infection.

Of the 3 billion prescriptions filled each year, 150 million are filled incorrectly. Two hospital patients out of every 100 (660,000) suffer adverse drug reactions resulting in increased length of stay (LOS) and $4,000 in additional costs. An estimated 7,000 patients will die as a result of a medication error. In addition, 2.5 percent of hospitalized patients suffer preventable adverse events, and 1,500 surgical patients per year suffer from "left ins" such as sponges or instruments. A 2003 *New England Journal of Medicine* study estimated that this happens once in every 1,000 surgeries. One in five orthopedic surgeons will conduct a wrong-site surgery during his or her career.

One New England hospital had *three* wrong-side brain surgeries in one year. Last year, a doctor failed to diagnose my mother's colon cancer for four months. My wife's sister has had extended complications from a perforated colon caused by a colonoscopy. Most healthcare workers tell me that medical mistakes are *underreported by a factor of 2 or 4.*

The Institute for Healthcare Improvement says that 50 patients out of every 100 will suffer some form of preventable "harm" while hospitalized. That's over 17 million patients a year.

And patients aren't the only people affected by healthcare problems. It is reported that 50 percent of hospitals have financial difficulties. Between 1986 and 1989, financial distress forced 231 acute-care hospitals to close, whereas 70 percent of rural hospitals and 50 percent of urban institutions were fighting to stay afloat. With 70 to 80 hospitals expected to close each year through the 1990s, the outlook was grim. The *Journal of Healthcare Management* reported that between 2000 and 2006, 42 U.S. acute-care hospitals filed for bankruptcy protection under federal law, and 67 percent of those hospitals *stopped operating.* The Internet even has articles about how to file for bankruptcy as a hospital. How can an industry producing 18 percent of the gross domestic product (GDP) be so sick?

Moreover, the costs of healthcare don't only impact the industry; they threaten the livelihood of patients and their families. From a patient's perspective, medical bills were involved in 60 percent of personal bankruptcy cases in 2007. The *Portland Tribune* reported in 2008 that Oregon hospitals wrote off nearly $450 million in debts from bills that patients were unable to pay. The high cost of healthcare is pushing patients into bankruptcy, and poor management of costs and receivables is pushing hospitals into bankruptcy. Lean Six Sigma can lower costs for both hospitals and patients while delivering superior profits and performance.

Having consulted with many hospitals and healthcare systems, I have seen that the common problems of patient flow, clinical mistakes, and operational errors are the same across the industry. Fortunately, Lean Six Sigma can solve these problems. The hard part is getting the hospital culture to adopt the improvements of Lean Six Sigma and sustain them.

A 2009 American Society for Quality study of 77 hospitals found that

- 53 percent use some form of Lean.
- Only 4 percent had fully implemented Lean.
- 42 percent had some form of Six Sigma.
- Only 8 percent had fully implemented Six Sigma.
- 11 percent were *unfamiliar* with either Lean or Six Sigma.

Most healthcare systems have gone through a number of implementations of *process improvement* (PI) or *total quality management* (TQM). While most healthcare workers have been dipped in some method for quality improvement, few have applied it successfully. This book is about *applied* Lean Six Sigma, not *theoretical* Lean Six Sigma.

Sister Mary Jean Ryan of SSM Healthcare says that with Lean Six Sigma, it's possible to achieve "breathtakingly better healthcare." She also says that it takes "superhuman tenacity" to make Lean Six Sigma part of a hospital culture.

TWENTY-FIRST CENTURY HEALTHCARE CHALLENGES

With healthcare reform all the buzz in Washington, most healthcare workers are hoping that they can just keep on going about their business and that the government will find a way to pay for universal healthcare for tens of millions of uninsured patients. To make healthcare affordable for all, however, healthcare will have to contribute dramatic cost savings.

Healthcare costs in America are over $2.5 trillion (Figure 1) and are expected to climb to $3.1 trillion by 2012, according to the National Coalition on Healthcare. Two and a half trillion dollars is a stack of $1,000 bills towering 168 miles high—essentially what the space administration considers low-earth orbit. The costs of waste and rework in any business, including healthcare, will range from 25 to 40 percent of the total ($800 billion to $1 trillion, says Paul O'Neill, former U.S. Treasury Secretary).

FIGURE 1

Healthcare costs as a percent of GDP.

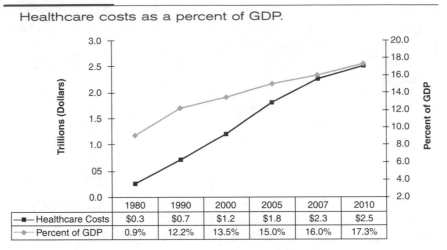

	1980	1990	2000	2005	2007	2010
Healthcare Costs	$0.3	$0.7	$1.2	$1.8	$2.3	$2.5
Percent of GDP	0.9%	12.2%	13.5%	15.0%	16.0%	17.3%

> A staggering 50 percent of healthcare consumed seems
> to be driven by physician and hospital supply, not
> patient need or demand.
>
> —CLAYTON M. CHRISTENSEN

"The U.S. healthcare system wastes $700 billion annually on the kinds of systemic inefficiencies that would make a quality management guru cringe," says Robert Kelly, vice president of healthcare analytics at Thomson Reuters. "About one-third of the country's total healthcare spending may be for unnecessary treatments, medical errors, redundant tests, administrative inefficiencies and fraud." And that's just the cost to the healthcare industry; it doesn't include the cost to patients, their families, and society, which is perhaps 10 times higher ($10 trillion).

Hospitals represent almost a third of healthcare costs (Figure 2). Hospitals are expected to deliver $113 billion of the $196 billion in savings mandated by the healthcare reform bill. Each of the nation's hospitals must cut $2.6 million a year.

STAGGERING STATISTICS

- One patient out of 100 admitted to a hospital will die as a result of a medical mistake (*To Err Is Human*, National Academy Press, 2000). With over 41 million admissions,

FIGURE 2

Healthcare spending.

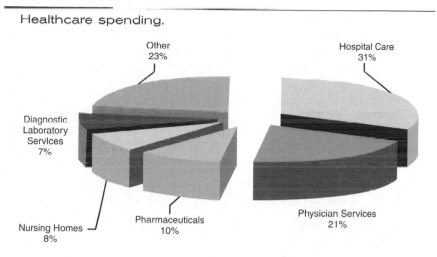

Other
23%

Hospital Care
31%

Diagnostic
Laboratory
Services
7%

Nursing Homes
8%

Pharmaceuticals
10%

Physician Services
21%

this means that there could be 410,000 preventable deaths each year, making healthcare one of the top 10 causes of death in the United States. The 2008 National Healthcare Quality Report found "that patient safety has actually been getting worse instead of better. . . . one in seven Medicare patients experiences one or more adverse events, and thousands of patients develop central-line-associated bloodstream infections each year." While the Institute for Healthcare Improvement's (IHI's) 100,000 Lives campaign hoped to reduce the preventable mortality rate, John Eisenberg, late director of the Agency for Healthcare Research and Quality (AHRQ), likened the problem of medical errors to an epidemic. A 2009 *Joint Commission Journal* article stated that healthcare doesn't know if the number of preventable deaths has actually decreased because healthcare doesn't measure them.

- Another six out of 100 patients will be permanently disabled as a result of a medical mistake.
- The IHI found that one patient out of every two (50 percent) suffered some form of *preventable* harm in the hospital. IHI's 5 Million Lives campaign hopes to reduce this.
- Fifteen out of every 100 diagnoses are incorrect.
- Twenty to 50 out of 100 diagnostic procedures (e.g., imaging) should never have been done because their results did not help to diagnose ailments or treat patients.

- Five to 10 out of every 100 admitted patients acquire an infection.
- Three patients out of every 100 have an incorrect ID band.
- The average hospital spends *one-quarter* of its budget on billing and administration.

Get the idea? Healthcare is rife with possibilities for improvement. Lean Six Sigma can help to minimize or eliminate all these problems. But healthcare clinicians and operational personnel have got to want to face the issues head-on and change the way healthcare is delivered to ensure better patient outcomes, faster service, and a better all-around experience.

The Institute of Medicine (IOM) set out six achievable goals:

- Patient safety—error-free care
- Effective treatment
- Patient-centered care (rather than physician-centered care) that optimizes the *patient's experience*, not the clinician's time
- Timeliness—reducing wait times and dangerous delays using Lean
- Efficiency—removing "waste" from the process using Lean
- Equitable care

HOW CAN LEAN SIX SIGMA HELP?

Fifty percent of a nurse's time is spent doing things that don't add value, like looking for medications that aren't where they're supposed to be or looking for equipment that isn't where it's supposed to be.

—PAUL O'NEILL,

Former U.S. Secretary of the Treasury,

January, 2001–December, 2002

While most visits to the emergency room (ER; hospitals call it the Emergency Department) take two hours or more, Robert Wood Johnson Hospital in New Jersey (Baldrige Award winner in 2004) does it in 38 minutes, on average. The hospital offers a 30-minute door-to-doctor guarantee. The hospital did it by rethinking the emergency experience from the patient's point of view. For busy soccer moms with a sick kid, the choice between ERs is easy; they can save 80 minutes by going to the fastest ER.

And the clinical side isn't the only issue. Healthcare operations—billing, ordering, etc.—waste even more money. Insurance companies are quick to reject claims and slow to pay, which causes more problems. I worked with one company that found ways to reduce denied claims from a single insurer and to start reaping an extra $380,000 a month.

The solution is simple. Most healthcare processes are set up like an assembly line, where the patient moves from station to station for treatment. The methods of Lean manufacturing can be applied to healthcare to get a *faster* hospital in five days. Rule 1: *Walking is waste.* When medical facilities are redesigned to prevent unnecessary movement of clinicians and patients, patient flow improves, and patient outcomes improve. When you stop watching the doctors or nurses and start watching how long the patient is waiting for the next step in his or her care, it's easy to see how to improve the process. But doctors and nurses have to be willing to trade 100 years of assembly-line tradition for dramatic improvement.

It's easy to get a *better* hospital (i.e., safer, less error-prone) in five days using a few key tools from Six Sigma. When you count and categorize the medical mistakes and operational errors, it becomes easy to see where the process or procedures are broken and how to fix them. And problems aren't spread all over like butter on bread; they cluster in a few key areas.

Rule 2: By focusing on the 4 percent of the healthcare that causes 50 percent of the delay, defects, and deviations, healthcare easily could boost quality, cut costs, and increase profits without breaking a sweat. But everyone from the CEO to the cleaning crew has to be on board with the effort. America voted for *change*, but I have found that most people want someone else to change; *they* don't want to have to change.

The good news is that any healthcare company can start today; the bad news is that they will never be finished. With focused effort, healthcare easily could save $1 trillion by 2012.

Lean Six Sigma is, first of all, a *mind-set* for solving specific problems surrounding the three demons of quality: delay, defects, and deviation. From a healthcare perspective, this means patient flow, turnaround times, medication errors, bloodstream infections, etc.

Second, it's easy to apply the methods and tools of Lean Six Sigma. Using Post-it Notes, it's easy to diagnose unnecessary delays in patient diagnosis, treatment, and flow. Post-it Notes make it easy to diagnose delays in operational processes as well. Using Microsoft Excel, it's easy to diagnose the root causes of mistakes, errors, and defects in clinical and operational processes.

Now for the bad news: Most businesses, while profitable, are barely at a 3 sigma (6 percent error rate) in performance. And this includes healthcare.

And now the good news: A handful of tools will take you from 3 to 5 sigma (0.03 percent error rate) in as few as 24 months. Yes, there is some statistical stuff, but it's easily handled by simple software that you can download *free* for 90 days. The QI Macros Lean Six Sigma software is an add-on for Microsoft Excel that is so easy to use that most people say that they can learn it in about five minutes. Forget all the fancy formulas. The QI Macros will do that for you. Just focus on what the graphs are telling you about how to improve the hospital. Since most business data are already kept in Excel or easily exported to Excel, you can get started using the tools right away. Without software, Six Sigma becomes too laborious for even the smartest employee, so the QI Macros will make it easy. You can download your 90-day trial from www.qimacros.com/hospitalbook.html. There are even free monthly Webinars and Money Belt Training on Lean Six Sigma and the QI Macros at www.qimacros.com/webinars/webinar-dates.html.

PIONEERS IN HEALTHCARE

> In health and medical care, it takes enough demonstration of performance close to [zero errors] in enough places of different sizes and scales across the country, so that it's undeniable that we can achieve much better results.
>
> —PAUL O'NEILL

Pioneering hospitals such as Virginia Mason, Cleveland Clinic, Mayo Clinic, and many others have started using the Toyota Production System (a.k.a., Lean) and Six Sigma to make dramatic improvements in all aspects of healthcare. SSM Health Care in St. Louis and the Pittsburgh Regional Healthcare Initiative have used Lean Six Sigma to

- Reduce coronary bypass readmissions by 4.7 percent, saving $1.7 million

- Reduce coronary bypass mortality by 25 percent "using a 3-cent aspirin and a 50-cent beta blocker in the emergency room"
- Reduce hospital-acquired infections by 85 percent, saving $30,000 per infection
- Reduce central-line infections by 63 percent
- Reduce medication errors from 16 per 100,000 to 1 per 100,000

Get the idea? Dramatic improvements are possible; most are simple, inexpensive, and good for patients as well as the bottom line.

CULTURE AND IMPLEMENTATION

The mind-set, methods, and tools of Lean Six Sigma are simple and easy to learn. Getting your healthcare culture to adopt these methods, tools, and mind-sets is the real challenge. If your employees are like most employees, you've experienced too many panaceas and programs-of-the-month. Your hospital probably tried process improvement off and on all through the 1990s. It's hard to keep Lean Six Sigma from ending up in the junkyard of failed culture changes.

Most Lean Six Sigma books and programs dive into the top-down endless training required to implement Lean Six Sigma. I call this the *wall-to-wall, floor-to-ceiling* approach to Lean Six Sigma. Unfortunately, research has shown that this method fails at least half the time. There are better ways to use the tools of Lean Six Sigma to get the results and culture change so often needed.

The best way, I have found, is to focus key personnel on making healthcare faster, better, and cheaper using only a handful of tools from the Lean Six Sigma toolkit. A pad of Post-it Notes and a flipchart can help you to identify ways to cut turnaround times by 50 percent or more from the emergency department to discharge. Control charts and Pareto charts will take any facility from 3 to 5 sigma (3 percent error to 0.03 percent error in 18 to 24 months). Most of these improvements can be discovered in *four to eight hours*. It may take a few days to implement the changes, but the solutions can be discovered quickly.

> **Hint: You don't have to know everything about Lean Six Sigma to do anything. A handful of tools will solve most of a hospital's problems.**

Healthcare employees want to do a good job. Give them an experience of using the methods and tools to make big improvements in the speed and quality of care, and they will embrace it. People learn best by doing. Burden them with week-long trainings and you lose them. I've worked with many teams in healthcare, and here's what I have learned:

> **Reality: It doesn't have to take weeks or months to make an improvement; the right team can figure out what to do and how to implement it in a matter of hours.**

This is why this book will focus on simple steps to start getting faster, better, *and* cheaper in five days or less while achieving the goal of fast, affordable, and flawless healthcare.

Jay Arthur

Simple Steps to
a Faster Hospital

With all the hoopla about healthcare reform, there's one huge missing piece—healthcare is going to have to get dramatically faster, better, and cheaper to help pay for the changes. Each of the nation's 5,700+ hospitals must find ways to cut millions of dollars in unnecessary costs over the next decade. This may sound difficult considering that half of all hospitals lose money. Most hospitals exist on a 4 to 5 percent margin. But Lean can help hospitals start getting faster, better, and cheaper in just a few days.

One of the key principles of Lean thinking is to eliminate delays that consume up to 95 percent of the total cycle time (57 minutes per hour). If you've ever been a patient in a hospital emergency room (ER) or nursing unit bed, you know that there are lots of delays. Over the years, healthcare has made tremendous strides in reducing cycle time in various aspects of care. Outpatient surgeries are one example: Arrive in the morning, and leave in the afternoon. No bed required. But there is still lots of room for improvement.

GOAL: ACCELERATE THE PATIENT'S EXPERIENCE OF HEALTHCARE

In any given "factory," there are two kinds of time: work time during a process when actual work is occurring and elapsed time—the total time a process takes (work time plus any time spent on handoffs, waiting, batches, backlog, and so on).

—KEN MILLER

Over the last decade, I've consulted with many hospitals on all kinds of projects. Perhaps the most powerful tool that can be applied immediately to start slashing cycle times, medical mistakes, and cost is Lean. And it doesn't have to take weeks, months, or years. With the right focus and the right people in the room, it only takes a few days to find ways to speed up any healthcare process, which, in turn, will reduce errors and boost profits.

Every hospital seems to have the same problem: *patient flow*. This shows up in many ways:

- Patient dissatisfaction and physician dissatisfaction
- Emergency Department (ED) divert hours (ambulances diverted because of overcrowding), patient *boarding* in the ED, LWOBS (leaving without being seen), and four-hour turnaround times
- Operating room (OR) delays, cancellations, and long turn-around times
- Imaging delays, long turnaround times
- Lab delays, long turnaround times
- Bed management delays
- Late discharges
- Long patient lengths of stay (LOSs)
- Lost revenue

What one element is critical to both patient flow and satisfaction?

Time—patient wait time leading to poor turnaround times and poor patient outcomes. Patient wait times are non-value-added (NVA). This non-value-added wait time is "baked into" existing procedures and facility design, but that doesn't mean that it can't be changed.

Healthcare delivery often involves complex processes that have evolved over time and that are neither patient-focused nor clinician-friendly. When systems do not work well, healthcare workers resort to creating "workarounds," adding additional layers of "patches" and "fixes" to poorly functioning systems.

—CHRISTOPHER S. KIM, M.D.

A FASTER EMERGENCY DEPARTMENT IN FIVE DAYS

In 2009, Press Ganey found that ED turnaround times still average *over four hours*, basically unchanged over the last decade. In 2006, the Centers for Disease Control and Prevention (CDC) found that 40 percent of hospital EDs were overcrowded. One Harvard study found that ED wait times rose 36 percent from 1997 to 2004.

Robert Wood Johnson Hospital in Hamilton, New Jersey (RJW), winner of the 2004 Baldrige Award, receives 50,000 patients a year. In 2004, RJW had ED turnaround times of

- 38 minutes for discharged patients
- 90 minutes for admitted patients

How is this possible? How did they do it? *By systematically eliminating the delays between registration, triage, examination, lab, imaging, and discharge or admission/transport.*

RJW's 15/30 Program

In 1998, RJW offered a 15-minute door-to-nurse and a 30-minute door-to-doctor guarantee. Like Domino's Pizza, if your nurse or doctor is late, your service is free! Patient satisfaction with the ED rose from 85 percent in 2001 to 90 percent in 2004. While payouts for this policy have been less than 1 percent of ED patients, ED visits doubled! This means that for 1 percent of the revenue, RJW increased ED revenue by 99 percent.

And because ED visits doubled, hospital revenue increased as well. Seventy percent of hospital admissions come from the ED. Faster turnaround times enabled the hospital to grow by over 10 percent per year, requiring the addition of a new nursing wing.

Results

- RJW was New Jersey's fastest-growing hospital from 1999 to 2003.

- Mortality rates for patients with congestive heart failure (CHF) declined from 8 to 2.5 percent.
- Infection rates for things such as ventilator-assisted pneumonia (VAP) fell from 10 per 1,000 vent-days to only 2 per 1,000 vent-days.
- The cardiology market share rose from 20 to 30 percent.
- The surgery market share rose from 17 to 30 percent.
- Hospital occupancy rose from 70 to 90 percent.
- Employee satisfaction with benefits rose from 30 to 90 percent.
- Employee satisfaction with participation in decision making rose from 30 to 90 percent.
- Retention of nurses rose to 98 percent.
- Employee retention rose from 80 to 98 percent.

Faster patient flow means greater patient satisfaction, better outcomes, and more money!

Studies have shown that patient satisfaction begins to decrease when ED LOS exceeds two hours. There are two populations of patients who visit the ED, so let's separate the emergent from the nonemergent cases and look at patients who get discharged.

If it takes only a couple of minutes to see the triage nurse, a few more minutes to get registered, and a few more minutes for a doctor diagnosis, then the total time spent on any one patient is perhaps nine minutes. So why does it take most EDs over two hours to handle each patient? Sure, some patients need lab work (11 minutes) and others need radiology, but most of those tests take less than an hour. We're still looking at 35 to 60 minutes, not two hours or more.

If we look at admitted patients, they are taken into the ED immediately without having to wait. They see the doctor immediately. Tests are done STAT. Registrations are done at the bedside. Nursing floor bed assignments take only a few minutes. Nursing reports are fast. Transport to the intensive care unit (ICU), cardiac care, or Med/Surg floor takes only 15 to 20 minutes. These patients should fly through the ED, but they take longer than the discharged patients—two to three times longer. Sure, they have to be stabilized, but why does it take hours to get them into an assigned bed?

The answer, across the board, is delay. There is too much time *between* activities. The admission staff is busy, so patients have to wait. The triage nurse is busy, so patients have to wait. The ED is boarding patients who should be in a nursing unit, so patients have to wait. The ED nurse can't reach the floor nurse to give a report, and vice versa. Neither nurse can leave to transport the patient. Beds are available but not staffed. And so on.

Imagine a Faster ED

Imagine an emergency room where patients walk in and something surprising happens:

1. They use the magnetic strip on their driver's license, insurance card, or credit card to check in and register using a kiosk. The kiosk automatically takes pictures of all these IDs and uses the data to find the patient's medical history, validate insurance, and so on.
2. Completing registration this way triggers a "pull" signal that brings the next nurse in the queue to collect the patient from the entry area and move him or her to an exam room.
3. Entering the exam room and gathering the patient's vital signs triggers a pull signal for the next ED doctor in the rotation.
4. The doctor examines the patient with the nurse available and requests any tests or x-rays using a handheld device that kicks off the orders.
   a. The nurse draws any blood or other samples required and either (1) sends them to the lab for processing or (2) uses point-of-care testing to get results in 11 minutes or less. (Approximately 70 percent of patients require lab work.)
   b. The nurse transports the patient to imaging, if needed. (Approximately 30 percent of patients require medical imaging.)
5. Completion of the tests triggers a pull signal to the ED doctor to collect the results, diagnose, and recommend treatment.
6. The doctor then initiates treatment. Any "teaching" material or paperwork required is prepackaged and ready for the nurse to prepare the patient for discharge or admission.

7. Initiating admission kicks off a pull signal for a bed in the appropriate unit. If there isn't enough staff in that unit to handle the admission, a pull signal may request an on-call nurse to come to work.

8. Instead of all being done manually, as most of this is now, it's all carefully orchestrated and technically linked to minimize delay. Many of these activities can happen in parallel, not sequentially as they do today.

A discharged patient is in and out in 30 minutes. An admitted patient is in a nursing unit bed in 60 minutes. Of course, there will be exceptions—a rush-hour accident may tie up one of the doctors—but most patients are discharged. Finding ways to handle them in "one-piece flow" will improve ED performance dramatically.

Simply speeding up discharge and housekeeping of nursing unit beds can alleviate boarding and overcrowding in the ED. Empowering triage nurses to order x-rays for possible fractures without doctor involvement can accelerate diagnosis and treatment. Scheduling radiologists during the hours of highest trauma injuries (think rush hour and Friday/Saturday night) can accelerate ED throughput. Prepackaging common triage kits can accelerate treatment.

FASTER DOOR-TO-BALLOON (D2B OR DTB) TIME IN FIVE DAYS

The ED at UMass Memorial Healthcare reduced D2B time in cardiac catheterization patients from 180 minutes in 2004 to less than 60 minutes. To optimize D2B times, the hospital measured and optimized the four key steps: (1) door-to-electrocardiogram (ECG) completion, (2) data to diagnosis, (3) diagnosis to decision, and (4) decision to balloon. Door-to-ECG time fell to 1 to 2 minutes, which enabled the ED physician to stay in the room to diagnose, decide, and initiate a call-in to the surgical team. On-call teams were scheduled with at least one team member located within 20 minutes of the hospital. Valet parking of team cars cut five minutes off the D2B time. Electronic ECG transmission from ambulances to the ED removed additional delays, allowing patients to go directly to the cardiac catheterization lab, bypassing the ED and reducing D2B time to less than 50 minutes. These changes reduced acute myocardial infarction (AMI) mortality to 11.7 percent, significantly below the 16.6 percent national average.

Lessons learned from D2B times were applied to door-to-incision time for vascular surgery and door-to-diuretic times for congestive heart failure patients.

In 2002, 16 hospitals in Virginia improved 90-minute D2B (door-to-balloon) times from 37 to 75 percent and reduced catheter complications from 3 to 1.4 percent and PCI complications from 4.4 to 2.5 percent by 2006.

A FASTER OPERATING ROOM IN FIVE DAYS

Copenhagen University Hospital wanted to reduce the time between surgical operations. The improvement team found that too much time (60+ minutes) was spent

- Investigating whether the patient got the required information from the surgeon (10 minutes)
- Unpacking individual sterile disposables (30 minutes)
- Waiting for missing devices (five trips per surgery)
- Waiting for the patient to regain consciousness to be transferred to recovery (20 minutes)
- Waiting for transport to recovery (10 minutes)

With some basic analysis, the team implemented countermeasures to save 60 minutes:

- Surgeon draws an X on patient's wristband when the patient has been informed about the operation, allowing anesthesia to begin.
- Prepackaged sterile disposables replaced individual disposables, saving two nurses and 30 minutes.
- Standard checklists ensure that all materials are gathered before the operation starts.
- Anesthetic depth was adjusted so that the patient wakes up when the operation is finished.
- Hospital orderlies move patients to recovery immediately.

SSM Healthcare reduced operating room turnaround times from 30 minutes to 15.8 minutes.

FASTER MEDICAL IMAGING IN FIVE DAYS

North Shore University Hospital wanted to improve patient throughput on its CT scanners to decrease LOS and increase patient satisfaction. Average turnaround time (TAT) was 20.7 hours and varied from 8 to 34 hours. The target for improvement? 16 hours. Identified problem areas included

- Manual scheduling process leading to calls from nursing units
- Time-consuming prep and delivery of contrast medium
- CT tech travel to requisition printer (6,480 feet per day)
- Transporter availability and travel (432 feet per day)

After analysis of these various issues, the improvement team implemented several countermeasures:

- The requisition printer was relocated in between the two CT scanners, saving over 6,000 feet per day of unnecessary travel.
- Dedicated CT transporter was assigned.
- An Excel-based schedule was maintained in imaging and was viewable by all nursing units (this reduced phone calls and cancellations owing to improper patient prep or availability).
- Instead of a rigid schedule with no room for STAT orders, a "pull" system adjusted the patient transport and scan to accommodate just-in-time STAT scans.
- Contrast preparation was reassigned to the evening shift, refrigerated, and delivered during the transporter's morning run for inpatients.
- One CT scanner was dedicated to complex procedures, and the second was dedicated to routine high-volume procedures to maximize patient flow.
- Staffing was adjusted to demand.

Results

- Average TAT fell from 20.7 to 6.45 hours.
- 200 additional inpatient scans were done per month.
- 60 additional outpatient scans were done per month.
- This resulted in $375,000 in additional revenue.

- Cancellations owing to improper prep dropped from 30.6 to 22.7 percent.

At Newton-Wellesley Hospital, radiology turnaround stood at 45 minutes. After examining patient and technician flow, the hospital found that technicians spent too much time walking around. By redesigning the work flow, turnaround times fell to 25 minutes, making the planned addition of another $500,000 x-ray machine unnecessary.

Massachusetts General Hospital's proton beam facility was fully booked, or at least so the hospital thought—until it did a little analysis. By batching patients requiring anesthesia on the same day and scheduling an anesthesiologist for that day, throughput increased from 29 patients per day to 39 patients—a 33 percent increase.

A FASTER LAB IN FIVE DAYS

In a large hospital, the response time of returning lab results to the wards was reduced by a factor of 10 simply by reducing the size of the transfer batch. It turned out that the transfer batch was determined by the size of the transfer trays.

—BOAZ RONEN

One 2,400-square-foot hospital lab wanted to reduce turnaround times, which would reduce ED turnaround times and reduce LOS in the nursing units. Using pedometers, the hospital tracked lab worker travel time for a week. The hospital conducted what's known as a 5S *analysis* (i.e., sort, straighten, shine, standardize, and sustain) to clean the area of 10 years' worth of clutter (Figure 1.1) in four hours, then mapped the value stream (Figure 1.2) in two hours, and redesigned the workflow (Figure 1.3) in two hours.

Using Post-it Notes and a flipchart, the lab team was able to redesign the lab to reduce

- Staff movement by 54 percent (goal 30 percent)
- Floor space by 17 percent (goal 10 percent)
- Phlebotomist travel by 55 percent (21,096 feet to about 4 miles, ~1.5 Full Time Equivalent [FTE])

FIGURE 1.1

5S lab trash.

- Tech travel by 40 percent (2,304 feet, 0.15 FTE over three shifts)
- Sample travel by 55 percent (23,400 feet and 7 hours of delay per 24 hours)

Some changes could be implemented immediately; others required coordination to move machinery and recalibrate. The lab got a lot faster with less than two days of effort.

If you go to www.qimacros.com/webinars/webinar-dates .html, you can sign up for a free Lean Six Sigma for Healthcare Webinar to learn more about this Lean for hospital labs projects.

Denver Health rearranged workstations and equipment in the lab to reduce turnaround times for test results by 25 percent and saved $88,000. Overall, since starting its Lean journey in 2005, Denver Health has saved $54 million in increased revenue and cost savings.

FIGURE 1.2

Lab original spaghetti diagram.

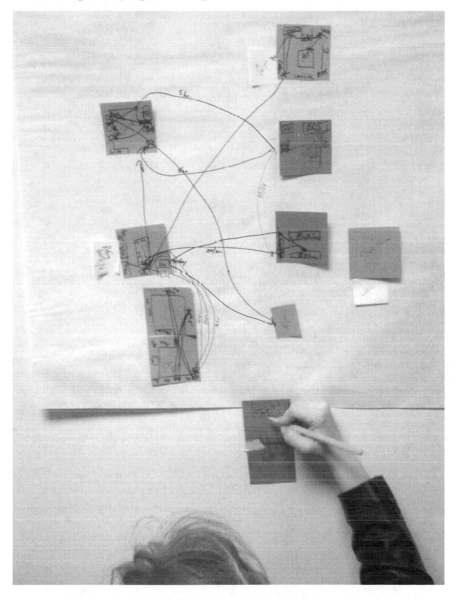

FIGURE 1.3

Lab redesign.

A FASTER NURSING UNIT IN FIVE DAYS

The same is true in nursing units. Nurses have to walk too far to get what they need. One redesigned nursing unit cut travel time by 67 percent, resulting in improved patient satisfaction, nursing satisfaction, and clinical outcomes. The unit got faster in a matter of days.

Additionally, nurses hesitate to take patients before shift changes, doctors make rounds at different times, orders are issued but not executed for a period of time, patients are discharged but no family member can collect them, and on it goes. Delay, delay, delay.

> ### The solution to this problem?
>
> ### Eliminate the delay.

THE PROBLEM ISN'T WHERE YOU THINK IT IS

Every department—ED, ICU, Med/Surg nursing floors, radiology, lab, housekeeping, bed management, and so on—thinks that it is doing the best job it can. Everyone is working hard, everyone wants to do a good job, everyone wants to serve the patient, but . . .

Insight 1: The Patient Is Idle Most of the Time

Rule 1: Stop watching your clinical staff. Start watching the patient, because the patient is idle 57 minutes out of every hour of the total turnaround time.

Patient LOS stay doesn't increase all at once. It increases in 10-, 15-, and 30-minute intervals.

- Why? Because the patient is idle, waiting for the next step in his or her diagnosis or treatment.
- Why is the patient waiting? Because the steps in his or her care haven't been linked up to eliminate delays.
- Why haven't the steps been linked up? Because no one is measuring and monitoring the time between steps.

Everybody seems to know how long it takes to do their job. It takes bed management 5 to 10 minutes to assign a bed, assuming that one is available. It takes 15 to 20 minutes to transport a patient to a bed. It takes housekeeping 22 minutes to clean an ICU or Med/Surg bed. The ED triage nurse takes only a few minutes to evaluate the patient. The ED doctor takes only a few minutes to examine the patient.

But nobody knows how long the delays are between each of these steps.

> **Rule 2:** Start measuring the delays *between* steps in the patient's care because this is how LOS increases and patient satisfaction *decreases*.

Insight 2: Walking Is Waste!

Any amount of time that a doctor, nurse, or technician spends walking is waste. Reduce the distance they travel, and it will improve patient satisfaction and outcomes.

Insight 3: Speed Is Critical to Patient Satisfaction!

Unfortunately, current hospital management practices discourage accelerating patient flow. The staff worries that if you move patients too quickly, they might have to send nurses home because of empty beds. Nurses depend on their income just like the rest of us, so they think they are actually being punished if they reduce patient delays.

As Robert Wood Johnson Hospital discovered, however, *faster patient flow leads to more jobs, not fewer*. Patients are smart; they can tell a faster hospital from a slower one.

Some members of the clinical staff think that accelerating patient flow means making the clinicians work faster or harder. But accelerating patient flow has little to do with clinicians; it has to do with reorganizing the work to get faster *patients*.

The clinical staff also worries that "haste makes waste," that faster turnaround times will lead to poorer outcomes, but this is true only if the clinician hurries. Accelerating patient flow isn't about making clinicians faster; it focuses on speeding up the patient. *Reducing delays between steps in patient treatment actually will give the clinician more time with the patient, not less.*

When the patient is handled in one, seamless interaction, there is less time spent learning what happened in the previous step (e.g., reading the chart) and more time spent with the patient. *Result: Improved patient satisfaction.*

Handling a patient seamlessly also prevents the opportunity to miss a step or do a step twice. Simply reducing delays will cut errors by 50 percent. *Result: Fewer medical errors and better patient outcomes.*

TAKE THE DOMINO'S CHALLENGE

Domino's made the guarantee that it could cook a pizza and deliver it to your home in 30 minutes or it was free. This began a revolutionary shift in customer expectations. Google taught everyone that it could find anything you want immediately and often for *free*. Customers used to want better, faster, and cheaper products and services; now they want everything free, perfect, and *now*, including healthcare.

This shift in customer expectations is hitting hospitals as well. If Robert Wood Johnson Hospital can offer a door-to-doctor *guarantee*, you might consider setting the same kinds of objectives:

- 30 minutes from door to doctor in the ED
- 30 minutes from "bed requested" to patient in bed
- 30 minutes from routine lab/radiology order to execution
- 30 minutes from discharge order to patient discharged
- 30 minutes from dirty room to clean room, dirty OR to clean OR

HOW TO GET A FASTER HOSPITAL IN FIVE DAYS

Although the case studies in this book offer some constructive ideas, most clinical staffs will not implement an improvement *unless they have a hand in its design.*

Improvements Are Possible If It Helps the Patient or the Provider

Healthcare professionals want to help create improvements that

- Increase patient safety and satisfaction

- Improve quality of care
- Reduce lead or turnaround times
- Improve productivity without compromising patient outcomes
- Reduce medical errors

How Is It Possible to Get a Faster Hospital in Five Days or Less?

1. Gather a team that believes that it is possible to improve patient flow (e.g., ED doctor, ED nurses, ED clerk, and ED admissions). Some people just don't believe that it's possible; if so, they won't be useful on the team. Don't load the team with skeptics.
2. *Prework:* Use pedometers to gather travel data about the clinicians. Identify and collect "wait times" for patients between steps in treatment.
3. Have a trained facilitator assist the team in identifying the major delays and unnecessary movement of people or supplies using tools such as value-stream mapping and spaghetti diagramming. Have the team identify possible countermeasures to these problems.
4. Implement the countermeasures, and measure results.
   - Implement process-oriented improvements immediately.
   - Move machines or supplies to more convenient locations immediately.
   - Project-manage more complicated changes (e.g., information technology changes, hardware changes, etc.).
5. Verify that the countermeasures actually reduce turnaround times. (Sometimes they don't.)
6. Standardize the improved methods and procedures as a permanent way of doing things.
7. Measure and monitor turnaround times to ensure peak performance.

Simple Steps to Faster Healthcare

It doesn't have to take forever to start making dramatic improvements. With a handful of Post-it Notes and a willing team, it's possible to speed up processes by 50 to 90 percent in a matter of days because all you're doing is eliminating unnecessary delays.

> **Faster care means better outcomes and fewer mistakes. Isn't it time to start accelerating the patient's experience to maximize his or her satisfaction?**

CHAPTER 2

Lean for Accelerated Patient Flow

We can significantly increase bottleneck throughput in sales, marketing, development and operations without adding resources.

—BOAZ RONEN

Medical educators must begin teaching tomorrow's doctors to become much better at creating, improving and managing processes and systems.

—CLAYTON M. CHRISTENSEN

The agonizingly slow flow of patients through a hospital's units from admission to discharge can cause delays in treatment, medical errors, and poor outcomes. Lean can simplify, streamline, and optimize patient flows for optimal productivity, profitability, and patient outcomes, effectively *eliminating the bottlenecks without adding resources.*

If you've ever been to London and ridden on the famous Underground, you've probably seen signs: "Mind the Gap" (Figure 2.1). While the signs are designed to keep travelers from wrenching an ankle, I believe that the idea also applies to Lean thinking.

FIGURE 2.1

Mind the gap.

MIND THE GAP

Here's what I mean: Hold up your hand and spread your fingers wide apart. What do you see? Most likely you're first drawn to look at your fingers, not the gaps between them.

This is how most people look at process improvement—by looking at the people working, not at the *gaps between people*.

When you take your eyes off the employees and put your eyes on the patient, product, or service going through the process, you quickly discover that there are huge gaps between one step in the process and the next. You'll discover work products piling up between steps, which only creates more delay—a bigger gap.

Your turnaround-time problems are in the gaps, not the fingers. You can make your people work faster, but you'll find that this often makes you slower, not faster, because more work piles up between steps, widening the gap, not narrowing it.

VALUE-STREAM MAPPING

The other sign you often see in the London Underground is a tube map (Figure 2.2).

FIGURE 2.2

London Underground map.

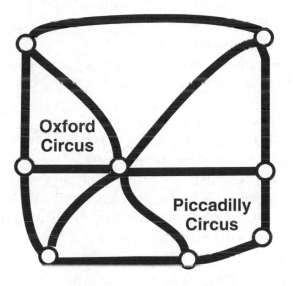

While it is much more interconnected than a typical value-stream map, you'll notice that the stations are quite small and the lines between them quite long.

This is true of most processes; the time between stations is much greater than the time actually spent in the stations. As this map suggests, 95 percent of the time is *between* stations, not in them. If you want to reduce the time it takes to serve a customer, you have to mind the gaps.

IF THEY CAN DO IT IN BOTSWANA . . .

At the American Society for Quality World Conference, I talked with two representatives from Botswana. They use the tools of Lean to *mind the gap* and reduce the response times of the police. They are using the benchmarks of the World Health Organization to improve healthcare response times. If they can do it in Botswana, then you can do it in your hospital.

Most hospitals have a blind spot when it comes to the best way to accelerate their response time. Stop trying to make the clinical and administrative staff faster. Stop trying to keep your people busy when there's no reason to be busy. Stop doing things patients haven't asked for yet. Focus on the time between steps.

If you want a faster hospital, *mind the gap*!

Recently, I took my wife to the Emergency Department (ED) at a local hospital because she was too dizzy to do anything. Here's the sequence of events:

1. I signed her in—and we waited.
2. The registration desk had us fill out some paperwork—and we waited.
3. The triage nurse took her vital signs and led us to an exam room—and we waited.
4. The doctor came in, listened to her symptoms, and ordered some tests—and we waited.
5. The phlebotomist took blood samples—and we waited.
6. The doctor came back and diagnosed the problem as an inner ear disorder and prescribed meclizine and bed rest.
7. Shirley got dressed and we left—two hours after we arrived.

In those two hours, we saw a clinician for perhaps six minutes total. The blood tests that actually take about 10 minutes to process took an hour of elapsed time. In essence, someone was working on diagnosis and treatment for only 18 minutes of the two hours.

Sadly, a typical ED is more like an old-fashioned assembly line, with patients going from processing station to processing station and piling up in between while they wait on the next operator. My 87-year-old mother describes her experience of healthcare as "feeling like a can of corn run over a scanner."

Just like customers in a fast-food restaurant, patients don't like waiting. According to Press Ganey, the average stay in an ED is *over four hours*, unchanged for decades. Doctors and nurses have gotten used to this sluggish process and assume that it's the best they can do. Not so!

Robert Wood Johnson Hospital (RWJ), which won the Baldrige Award for Quality in 2004, cut ED turnaround times dramatically: 38 minutes for a discharged patient and 90 minutes for admitted patients.

In 1999, RWJ started offering a 15–30 guarantee: See a nurse in 15 minutes and a doctor in 30 or your visit is free. Soon soccer moms and families figured out that they could get in and out of the ED *fast*. The hospital grew at over 10 percent per year and had to add a new wing.

THE FAST EAT THE SLOW

How did RWJ do what others consider impossible? The hospital eliminated the delay between steps. How is that possible, you might ask? There are lots of opportunities:

- Why sign in? Why not swipe the magnetic strip on your insurance card, driver's license, or credit card just like you do at an airport?
- Why have a registration desk when computers on wheels (COWs) can go to patients in an exam room to gather the information necessary while they're waiting on test results?
- Why photocopy a patient's driver's license and insurance card when lightweight scanners make it easy to do and attach to the patient record.
- Why have a triage nurse? Why not have every nurse in a rotation *pull* the next patient as he or she arrives?
- Could the doctor gather the patient's symptoms *while* the nurse takes the patient's vital signs? Why does everything have to be done serially rather than in parallel?
- Why send samples to the lab? Point-of-care testing is getting cheaper and easier. Moore's law says that computers double in power and halve in cost every 18 months. Soon the analysis needed will be at a clinician's fingertips.
- Is there a way to expedite admissions? Bed assignment? Of course!
- Can a nursing unit "bid" to take a patient from the ED based on the acuity and nurse-patient ratio in the unit?
- Can the nursing unit transport the patient if the ED is busy?

Let's face it, a hospital can't afford to have patients boarded in the ED. Boarding leads to diversion, which leads to lost revenue:

- Value per ambulance: $6,000 or more
- Ambulances per hour: two
- Heart surgery patients: one out of six ambulances ($100,000+)

Based on these kinds of numbers (plug in your own), three hours on divert will cost a hospital $130,000 and possibly cost a patient his or her life. While nursing units are trying to maximize their nurse-patient ratios to save a few bucks, the big money is diverted to other hospitals. The chances of diversion increase rapidly when the census (occupancy) exceeds 80 percent.

And what about LWOBSs (leaving without being seen)? How many people leave or don't even enter the ED if the waiting room looks too crowded? Answer: Lots!

The ED is the front door to the hospital, admitting one out of every four patients using the ED. If you really want to be in the patient care business, you will want to get a lot faster than you are.

What are some other examples where speed matters?

- Door-to-balloon (D2B) time under 90 minutes for cardiac catheterization patients
- Lab turnaround times
- Imaging turnaround times
- Discharge turnaround time
- Room cleaning
- Meal delivery

CORE SCORE

In Marcus Buckingham's book, *The One Thing You Need to Know*, there's a section on knowing your "core score." What's the one thing you need to know about your business?

Core Score for Prisons

Buckingham interviewed General Sir David Ramsbotham, who was in charge of Her Majesty's prisons. He says that he knew he couldn't make wardens change. In order to make things happen, he had to change the way they measured success:

- *Old metric:* Number of escapees
- *New metric:* Number of repeat offenders

The old goal was to keep prisoners in, but the general started thinking: Who is a prison designed to serve? Answer: The prisoner! "The main purpose of a prison should be to serve the prisoner. By which I mean that we must do something for the prisoner while he is in prison so that when he is released back into society, he is less likely to commit another crime." Armed with this new score, he turned the prison world upside-down.

Core Score for Healthcare

In the old world of healthcare, the measure was based on "outcomes"—Did the patient get better no matter how long it took? I am coming to believe that the new world of healthcare is measured in speed.

- Door-to-doctor time in the ED of less than 30 minutes
- ED length of stay (LOS) of under an hour
- ED-to–nursing floor for admitted patients of under 30 minutes
- LOS of two to three days based on diagnosis
- Discharge-to-disposition (patient transferred) of under 60 minutes

Most of these times can run two to four times longer at present. Patients are used to being served in minutes everywhere else. Why not in healthcare? Of course, healthcare will need a few metrics of patient safety as well.

Most people worry that faster treatment will result in worse outcomes, but if you use Lean correctly, the opposite is true.

SPEED SAVES LIVES

You may remember when the speed limits were lowered to 55 mph to "save lives." Yet a study by the Cato Institute found just the opposite: The fatality rate on the nation's roads declined for a 35-year period excluding the period from 1976 to 1980 when the *speed limit was 55 mph*. After the speed limit was raised in 1995, the fatality rate dropped to the lowest in recorded history. There were also 400,000 fewer injuries.

Furthermore, there's no evidence that states with higher speed limits had increased deaths. States with speed limits of 65 to 75 mph saw a 12 percent decline in fatalities. States with a 75 mph speed limit saw over a 20 percent decline in fatality rates. Similar improvements are possible in healthcare.

Patient Safety Alerts

Virginia Mason Medical Center (VMMC) in Seattle is widely regarded for its quest for patient safety and quality. Gary Kaplan, CEO, offers many insights into the hospital's journey. It all began in 2001 when the board began to ask, "What is best for our patients?"

As a Lean Six Sigma practitioner, I would say that healthcare normally has asked, "What's best for our physicians?" and then asked, "What's best for our patients?" A physician-centric model leads to unnecessary inefficiencies and errors. A patient-centric model changes the way healthcare is delivered.

VMMC began looking for quality leaders and, through Boeing, found the Toyota Production System (TPS)—Toyota's relentless focus on the customer, quality, and safety—and modeled it to create the Virginia Mason Production System (VMPS). Kaplan says, "When you find and eliminate defects, quality and safety are improved."

In 2002, VMMC leaders went to Japan to observe Toyota's production line firsthand. When they observed Toyota's "stop the line" concept, where any employee can slow or stop the line when he or she notices a defect and everyone rushes to "go and see" what's going on and fix it, it was a revelation: "If Toyota does this for cars, shouldn't we be doing this for our patients?"

"In conventional manufacturing they keep the line going and inspect products afterward. The way it is traditionally done in healthcare is even worse," Kaplan says. "Two months after the fact, a retrospective quality audit or chart review will be done and find that something should have been done differently—two months ago."

VMMC created a Patient Safety Alert (PSA) system "Every single staff member is a patient safety inspector and is empowered to stop any process or situation that might cause harm to a patient. To [2009], we've had more than 14,000 PSAs. While it may seem counterintuitive, we think the more PSAs we generate, the safer care is here." To encourage PSAs and achieve zero defects, VMMC has three key expectations:

- It's safe to report mistakes.
- When mistakes are reported, they will be corrected.
- Those who report mistakes will be praised.

When a PSA is reported, managers and staff use the "five whys" to diagnose and prevent the problem from recurring.

So what's wrong with the system? Kaplan says that "too many residents and nursing students are still taught that if you try hard enough, you won't make mistakes. . . . But we know that's not true. . . . humans make on average six errors a day." Think about it: Multiply the daily staffing level by six. How many thousands of errors occur every day? The only choice is to change the system to make those errors *impossible*.

"The healthcare industry has a defect rate of 3 percent or more," Kaplan says, "a rate unacceptable in any other industry.

Defects can be anything from appointment scheduling errors to wrong-site surgery. Zero is the only acceptable rate for defects. Perfect care must be our goal."

Cathie Furman, senior vice president of quality and compliance, asked, "What's the single easiest thing every hospital or clinic could do to make significant strides in patient safety?" Kaplan says, "Listen to staff . . . and get leaders out of their offices. *Genchi Genbutsu* means 'go and see for yourself.'"

Kaplan says, "If we reduce cost by reducing waste, which is what VMPS is all about, we actually improve quality." Most people find it counterintuitive, but when you speed up the process using Lean, it actually reduces defects by 50 percent or more because there are fewer chances for error.

Insight: Haste makes waste, but speed makes quality.

Could your hospital implement a PSA system that allows every employee, patient, or family member to call an alert that "stops the line" and allows for immediate improvement? It requires

- Executive leadership
- Easy, open reporting—"Open the floodgates for all concerns"
- Rapid change as the organization learns

It took VMMC from 2003 to 2004 to get PSAs up from 125 and 204 a year to 2,450 in 2005 and 3,315 in 2006. It takes time for employees to learn to trust that alerts will be dealt with and improvements incorporated into the system, especially the nurses and nonclinical support personnel, who contribute two-thirds of PSAs.

Lean principle: Eliminate unnecessary processing.

VMMC also focuses on appropriateness of care. If a patient doesn't need a procedure or treatment, then, no matter how well it was performed, "There is no quality." It happens all too often. Robert Kelley estimates that 40 percent of healthcare waste is due to *unnecessary care*. "Unwarranted treatment to protect against malpractice exposure accounts for $250 billion to $325 billion in annual healthcare spending. Twenty percent to 50 percent of [high-tech scans] should never have been done because their results did not help diagnose ailments or treat patients."

How do you maintain the gains and continue to improve? As Deming would say, "Constancy of purpose at all levels of the organization." Kaplan says, "The willingness to listen to the voice of the patient is what keeps us moving forward."

YOU ALREADY UNDERSTAND LEAN

I'd like to suggest that you already have been exposed to and understand the concepts behind Lean. Kitchens, for example, have long been designed as "lean cells" for food preparation. The refrigerator, sink, and stove should form a V-shaped work cell. The tighter the V, the less movement is required of the cook. My kitchen looks like Figure 2.3: Food comes out of the refrigerator, gets washed in the sink, cut up on the counter, cooked on the stove, and delivered to the dining table in a neighboring room. Unlike mass production, where different silos would be put in charge of frozen and refrigerated food, washing, cutting, and cooking, there's usually only one cook who handles each of these steps. Each meal is a small "batch" or "lot." You never cook in batches big enough for the entire week. A trip to the supermarket each week replenishes the limited inventories of raw materials required. Ever noticed how most kitchens are right off the garage? In this way, each week's groceries come straight out of the garage right into the kitchen with a minimum of movement. Your kitchen is the essence of a Lean production cell.

FIGURE 2.3

Lean kitchen layout.

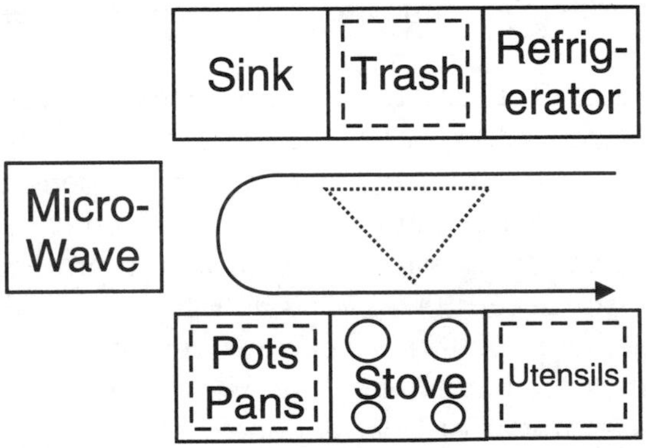

How can you set up your workplace (ED, operating room, lab, radiology suite, nursing unit, intensive care unit, administration) to use the insights gleaned from your kitchen?

THE POWER LAWS OF SPEED

It's not the big that eat the small, it's the fast that eat the slow.

—JASON JENNINGS AND LAURENCE HAUGHTON

The amount of time it takes to deliver a product or service is far greater than the actual time spent adding value to the product or service. Most patients, products, and services are worked on for only three minutes of every hour of the total elapsed time (value-added). Why does it take so long? Delay. The patient, product, or service is sitting idle for far too long between steps in the process (non-value-added).

Being busy is a form of laziness—lazy thinking and indiscriminate action.

—TIMOTHY FERRISS

The 3-57 Rule: Your employees are busy, but they are only working with each patient for three minutes out of every hour. Your patient is waiting for 57 minutes of every hour. *Watch your patient, product, or service, not your physicians, nurses, or technicians.*

The 15-2-20 Rule: Every time you reduce the time required to provide a product or service by 15 minutes per hour, *you double productivity and cut costs by 20 percent.* It has been my experience that you can reduce cycle and turnaround times by 30 to 50 minutes per hour so that your productivity increases, and cost reductions and profit margin improvement should far exceed the 15-2-20 rule.

The 3×2 Rule: When you slash your cycle time for your mission-critical processes (door to doctor, door to test, or door to balloon), you enjoy growth rates three times the industry average and twice the profit margins. This is a good thing because most companies find that they only need about two-thirds of the people to run the business after applying Lean, but if you're going to grow three times faster than the industry, you're going to need all of those employees to meet the demand.

ECONOMIES OF SPEED

There is always a best way of doing anything.

—EMERSON

One of the best ways to improve a process is to find and eliminate as much of the delay as possible. Although the people are busy, the customer's order is idle up to 95 percent of the time—sitting in queue, waiting for the next worker. Delays occur in three main ways:

- Delays *between* steps in a process
- Delays caused by waste and rework
- Delays caused by large batch sizes (The last item in the batch has to finish before the first item can go on to the next step. This is not a common problem in healthcare.)

So the obvious answer is to eliminate delays by

- Eliminating delays *between* steps
- Using Six Sigma to reduce or eliminate the defects and variation that cause waste and rework
- Reducing batch size (to one, if possible) (Toyota, for example, can produce up to nine different models of car on the same production line *simultaneously* and customize each one produced. In healthcare, the clinical batch size is one patient at a time. The paperwork and accounting batch size is often much larger.)

THE TOYOTA PRODUCTION SYSTEM

Toyota invented "lean production" according to Jeffrey Liker, author of *The Toyota Way*. It's also known as the *Toyota Production System*, or TPS for short. And it seems to work equally well in manufacturing and healthcare.

Here's Toyota's critical discovery: When you make lead times short and focus on keeping production lines flexible, you actually get better quality, responsiveness, productivity, and utilization of equipment and space.

CORE IDEAS OF LEAN

The principles of Lean are pretty simple, whether you apply them to manufacturing, healthcare, or administration:

- *Pull versus push* (Let the patient *pull* the product or service; don't push.)
- *No delay versus delay* (Eliminate delays.)
- *Parallel versus sequential* (Do more things in parallel.)
- *No action versus action* (Eliminate unnecessary processing.)
- *No movement versus movement* (Eliminate unnecessary movement.)

1. *Determine value using the voice of the customer.* What does the patient want? Determining value, from the customer's point of view, can be a challenge for a number of reasons:
   - Value is an *effect* of doing things right. The effect of improving speed, quality, and cost leads to higher patient satisfaction, retention, and referrals—all of which lead to growth and profitability.
   - What has value in one situation will not have value in another. If I want a colonoscopy on a Friday, it doesn't matter to me if you can do it on any of the weekdays *before* Friday. (I want it when I want it, not when you can deliver it.)

2. *Use "pull" systems to avoid overproduction.* Big inventories of raw materials or finished goods hide problems and inefficiencies. You may not think of hospitals as having big batches, but the nightly run of lab tests is a big batch. How can these big batches be reduced to a more manageable level?

3. *Institute one-piece flow.* Make the work *flow* so that there are no interruptions or wasted time or materials. Clinically, patients are treated one at a time, but more like a manufacturing production line than a race car pit stop.

4. *Level out the workload* (hejunka) *to the rate of customer demand or pull.* Instead of scheduling any surgery on any day, why not schedule the operating room (OR) by recovery time to optimize nursing units.

5. *Stop and fix problems immediately to get quality right the first time.*

6. *Standardize to support improvement.* Consider checklists as a way to standardize processes and instrument carts for various surgical procedures.

7. *Use visual controls so that no problems remain hidden.* Color-code supply bins to optimize nurse search times. A whiteboard "map" of the intensive care unit (ICU) or nursing rooms showing patient names and status gives an immediate overall view of the unit and its patients. VMMC uses colored wrist bands to signal conditions such as allergies, fall risk, and do not resuscitate (DNR) orders.

8. *Use only reliable technology that supports the people and the process.*

9. *Compete against perfection, not competitors.*

Toyota's Five Rules

At the Harvard Macy Institute program for academic healthcare leaders and medical school deans, deans arrive believing the explosion in knowledge students need to absorb means that medical school will need to be lengthened to five years. By the end of the course they realize, almost without exception, that by following [Toyota's] five rules, better doctors can indeed be trained in three years.

—CLAYTON M. CHRISTENSEN

In *The Innovator's Prescription*, Clayton M. Christensen reviews the four elements of a process and five rules Toyota has embedded in its DNA, as Steven Spear discovered years before. The four elements of a process are activities, connections, pathways, and improvements. The five rules are:

1. Each step in a process must be completely specified so that it can be done perfectly each time without subsequent rework.

2. Never work on a "part" (e.g., a patient) until it's ready to be used in the next step. When you do this, it immediately tests whether the prior step was done correctly.

3. The sequence of steps must be completely specified as one-to-one handoffs. "Any worker to any worker hand-offs are not allowed."

4. Perform each step the same way every time—not to make the work mindless, but to test if it is the best way to do it.

5. Never allow a problem to persist by working around it. Use root cause analysis to eliminate the problem.

If medical school deans can use these rules to redesign curriculums to turn out better doctors in three years instead of four or five, aren't they worthy of application to the entire hospital flow?

Christensen also points out that with today's technology, students don't need to arrive in the fall in a large group. They can arrive in small groups and "listen to the best lecturers in the world, anytime, anywhere at the click of a mouse." This adheres to the Lean concept of one-piece flow.

THE LEAN MINDSET

Here's the mind-set shift that you will want to embrace to understand Lean:

From: If you build it, they will come (mass production).

To: When they come, treat them fast (lean production).

What's weird about this kind of thinking?

1. The top priority is to treat patients at the rate of patient demand, *not to keep clinical workers busy.*

2. Sometimes the best thing you can do is to stop doing stuff.

3. Create only a small inventory to level out the production schedule (e.g., the nightly run of lab tests can send two or three samples at a time to the lab via pneumatic tube rather than collecting all the samples in a nursing unit and returning them, in bulk, to the lab for processing).

4. The more inventory you maintain, the less likely you will have what you need! Too much inventory creates clutter and hides shortages. Too much purified water in a lab creates a roadblock. Too many kinds of gloves hide a shortage.

5. It's usually best to work out a process manually first before adding technology.

THE SEVEN SPEED BUMPS OF LEAN

The seven speed bumps of Lean focus on non-value-added waste, which includes any activity that absorbs money, time, and people but creates no value. Toyota describes these as

1. *Overproduction (the most common type of waste), which creates inventories that take up space and capital.* One ED I visited took blood samples from *every patient* just "in case" they needed them. This is classic overproduction.

2. *Excess inventory caused by overproduction is waste.* If the ED, for example, processes patients faster than nursing units can absorb admissions, the patients end up "boarded" in the ED. Patients waiting in the ED waiting area are excess inventory. Patients waiting in an ED exam room for lab results are *work in process* (WIP), and WIP is a *problem, not an opportunity.*

3. *Waiting.* Don't you hate standing in line? So do your patients. Are they always waiting for the next value-adding process to start? Don't you hate waiting for your computer to boot up? So do employees. Are they waiting for missing supplies or late attendees? Waiting is waste.

4. *Unnecessary movement of patients (i.e., transportation).* When you break down the silos into cells, the patients and products don't have to travel as far between processes.

5. *Unnecessary movement of employees.* Are supplies and tools too far from where they're needed? Are employees walking too far to get supplies or deliver a work product? Most hospital labs, ORs, radiology suites, and nursing units can be redesigned to reduce travel by 40 to 50 percent, leaving more time for patients.

Insight: Walking is waste!

6. *Unnecessary or incorrect processing.* Why have people watch a machine that can be taught to monitor itself? (Toyota calls this "autonomation"—automation with a human touch.) Why do things that add no value? Is one group

doing something that the next group has to correct? Stop doing more than what the patient or customer wants, and start doing everything right the first time. Unnecessary medical testing costs $250 billion or more a year. How can we eliminate it?

Recently, my wife had to take her aunt to the ED. Her aunt had a CT scan near the end of a shift. After the shift change, a new nurse came in and asked my wife's aunt to drink another bottle of contrast material. Although her 82-year-old aunt was ready to follow the instruction, my wife intervened: "She's already had a CT scan." "No she hasn't," said the nurse. "Go check," my wife said. They finally got the confusion sorted out, but this almost resulted in unnecessary processing.

7. *Defects leading to repair, rework, or scrap.* Lean will help you to reduce or eliminate numbers 1 through 6. Six Sigma will help you to reduce number 7. When you rearrange your ED, OR, radiology suites, and nursing units into production cells with right-sized machines and quick changeover, you can quickly reduce most of these common kinds of non-value-added waste by 50 to 90 percent.

8. *Waste of talent and ideas.* Some Lean practitioners add this eighth waste. In a hospital, a surgeon waiting for an OR is a waste of talent. A nurse with an idea about how to reduce travel time in a nursing unit but no authority to implement it is a waste of both talent and ideas. Are you wasting the talent in your hospital because of poor process or workspace design?

Here are some simple questions to identify and remove waste:

Work time:	How long does this step take if you do it without interruptions?
Wait time:	How long does the patient or product wait for the next step?
Constraints:	What are the bottlenecks in this process?
Inspection:	Does this step just inspect the product or service, but not add value?

Rework: Does this step just fix previous errors or mistakes?

Batch size: Is the first work product waiting on the last one before moving on?

Inventory: Is there too large an inventory of materials, which causes confusion?

THE FIVE S'S

To remove the waste, we turn first to the five S's. The 5S concepts are a great way to really understand what's going on in your process. The 5S principles of reorganizing work so that it's simpler, more straightforward, and visually manageable are

1. *Sort.* Keep only what is needed. Pitch everything else.
2. *Straighten.* A place for everything, and everything in its place. Make everything visual for quick access.
3. *Shine.* Clean machines and work areas to expose problems.
4. *Standardize.* Develop systems and procedures to monitor conformance to the first three rules.
5. *Sustain.* Maintain the standard processes for sorting, straightening, and shining.

I've worked in many hospital laboratories. If the lab has been around for more than a few years, there's usually a lot of inventory to 5S. It takes about four hours to 5S a 2,000-square-foot lab. Lab workers usually find one to two dumpsters' worth of stuff to throw away. It's amazing how many chemicals are left over from prior equipment. There can be three places for the same pipette at one workstation instead of just one place. There are stashes of gloves all over the lab, not in one place, which causes overordering and shortages.

Once you've thrown away all the clutter and organized what remains, you can more easily see the products flowing through your workspace.

Red Tagging

Of course, you might be afraid of throwing away something important. Simply put a *red tag* on it showing the date discarded, and put it in a place designated as the "red tag room." In this way, other shifts can find and retrieve needed items. (This rarely happens.) At the end of 30 days, *throw it away or donate it to some cause.*

VALUE-STREAM MAPPING

Automation applied to an inefficient operation will magnify the inefficiency.

—BILL GATES

Having just done the five S's on your "factory," you'll be in great shape to understand the overall value stream. A key starting point for implementing Lean is the concept of *value* and the *value stream.* Value is defined by the customer (i.e., patient, family, or payer), not the company, business unit, manager, or employee. When I worked in information technologies, for example, programmers often focused on cool, new technology, not on what was fast, proven, and effective for the customer. Craftspeople bear allegiance to their craft, not to their customer.

Since most hospitals have grouped work together into functional silos, each silo often skews the definition of value. While each silo attempts to optimize its own operation, the hospital fails to optimize the overall flow of patients and services, which creates tremendous waste.

Become the Patient or Service

Most people find this hard to believe, but when you take the perspective of the patient or service and notice how long you sit around waiting for something to happen and how many things go wrong and have to be reworked, you get some idea of the waste in the process. All this delay and rework can be eliminated using Lean.

Whenever I work with a group on Lean, I start wherever the product, patient, or service starts and follow it around. I ask dumb questions about why things are done this way. The usual answer is that it's always been done this way. Then I'll ask: What if we move this machine over there so that the product or employee doesn't have to travel so far? Often the team will say it can be done. Then I ask: *Can we do it now?*

This is the essence of Lean. The moment you notice one of the seven speed bumps, ask yourself: Can I change this *now*? If so, just move the machine, tool, or material. Most people are surprised when Japanese counselors come into a manufacturing plant and just start moving machines into production cells. Don't study it to death; get on with making things better.

PULL VERSUS PUSH

Once you understand what the *customer* wants, then you can redesign the process to produce it in a way that minimizes time, defects, and cost. The secret is to only produce the product or deliver the service *when* the customer asks for it. This is the essence of a "pull" system. In a hospital, a patient walking into an ED for care or lab for testing is a *pull signal*. A doctor issuing a discharge order for a patient is a *pull signal* that should trigger action for patient instruction and transport. A discharged patient leaving a room should trigger a *pull signal* for housekeeping to clean the room. A clean room should trigger a *pull signal* for a new patient from bed management. Get the idea? Once you know the pull signals, it's time to redesign for one-piece flow.

REDESIGN FOR ONE-PIECE FLOW

What are the benefits of one-piece flow?

1. It builds in quality.
2. It creates flexibility.
3. It increases productivity.
4. It frees flow and space.
5. It improves safety.
6. It improves morale.
7. It reduces inventory.

Here's the mind-set shift for one-piece flow:

From: Big batches

To: Single pieces or small batches

The trick is eliminating all the delay between value-adding steps and lining up all the machines and processes so that the patient, product, or service flows through the value stream without interruption. Mass production and large batches ensure that the product will have to sit patiently waiting for the next step in the process. The mental shift required to move from mass production to Lean thinking is to focus on continuous flow of small lots.

Every time I look at a waiting room in a hospital or doctor's office, I think "big batch." It's designed to optimize the physician's time, not the patient's. This is a mistake. When you optimize the patient's time, you'll automatically optimize the clinician's time.

Common Measures of Flow

- *Lead (or cycle) time:* The time product stays in the system
- *Value-added ratio:* Value-added time/lead time
- *Travel distance of the product or people doing the work*
- *Productivity:* People hours/unit
- *Number of handoffs*
- *Quality rate or first-pass yield*

The Redesign Process

1. The first step is to focus on the patient, product, or service itself. Follow the patient through the entire treatment cycle. In a hospital, you would follow a patient through from admission to discharge. In a printing company, you'd follow a job from start to delivery. In a manufacturing plant, you'd follow the product from order to delivery. You can use a spaghetti diagram to show the movement of parts, products, and people through the current production maze.

2. The second step is to ignore traditional boundaries, layouts, etc. In other words, forget what you know about how to handle patients.

3. The third step is to realign the workflow into production *cells* to eliminate delay, rework, and scrap (e.g., OR or catheterization lab).

4. The fourth step is to *right-size* the machines and technology to support smaller lots, quick changeover, and one-piece flow. This may mean using smaller machines that actually may be more accurate and reliable. In a lab, it may mean

using a STAT centrifuge instead of a bucket centrifuge because it takes less time.

The goal of flow is to eliminate all delays, interruptions, and stoppages and not to rest until you succeed.

Cell Design

A *cell* is a group of workstations, machines, or equipment arranged such that a product can be worked progressively from one workstation to another without having to sit and wait for a batch to be completed and without additional handling between operations. Cells may be dedicated to a process, a subcomponent, or an entire product. Cells can be designed for administrative as well as manufacturing operations. An exam room in an ED is a work cell.

Cell design helps to process products with as little waste as possible. Arrange equipment and workstations in a sequence that supports a smooth flow of materials and components through the process with minimal transport or delay. Cells can help to make your company more competitive by

- Cutting costly transportation and delay
- Shortening lead times
- Saving floor space
- Reducing inventory
- Encouraging continuous improvement

A factory work cell contains three to nine people and workstations in a compact U-shaped arrangement (Figure 2.4). Cells ideally process a range of highly similar products. A cell should be self-contained with all necessary equipment and resources. The U shape makes communication easy because operators stay close to each other. This improves quality and speed. Can the hospital lab be redesigned to minimize travel and optimize turnaround times using this type of cell?

Many hospital EDs have their own portable x-ray machines; the patient doesn't have to move at all to be x-rayed. Some also have CT scanners and even MRI machines to reduce patient travel and accelerate diagnosis. Some EDs are using point-of-care (POC) lab testing. If you can get lab test results in 10 minutes in the ED versus 40 minutes in the lab, this shaves 30 minutes off your patient's wait time and accelerates flow through the ED. While the

FIGURE 2.4

Lean work cell design.

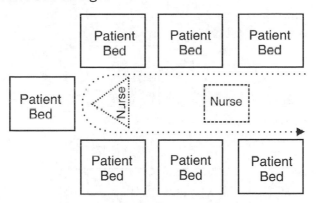

cost per test is currently higher, it also costs an estimated $6,000 or more to turn away an ambulance when the ED is full. POC testing may be cheap when you look at the overall cost.

LEAN TOOLS

Having said all this, there are two key tools used to help visualize problems with speed:

- *Value stream maps* (Figure 2.5)—to visualize the flow of the process
- *Spaghetti diagrams* (Figure 2.6)—to visualize the flow of *work* through the production area

With these two tools, you can identify 80 to 90 percent of all problems associated with delays and non-value-added waste.

If you go to www.qimacros.com/hospitalbook.html, you can download the QI Macros Lean Six Sigma Software 90-day trial. Click on the "Fill-in-the-Blank Lean Tools" to explore the value-stream mapping template for Excel.

A simple way to begin is to map the value stream and analyze each element for non-value-added waste and delay. Then redesign the flow to remove as much of the non-value-added waste as possible and standardize the ongoing process.

Value-stream mapping assumes that an idle resource is a wasted resource. An activity or step that doesn't in some way benefit a customer directly is also wasteful.

FIGURE 2.5

QI Macros value-stream map.

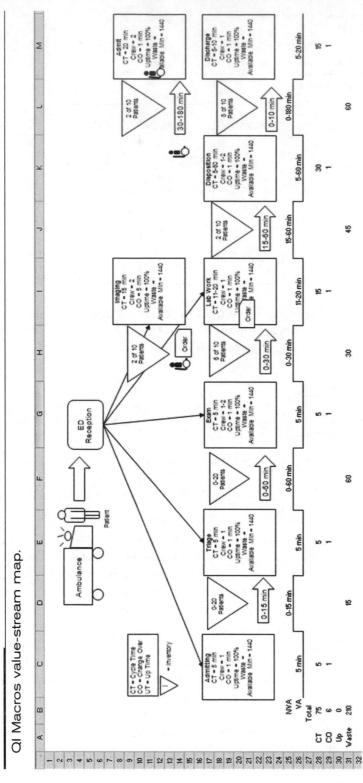

FIGURE 2.6

Hospital lab spaghetti diagram.

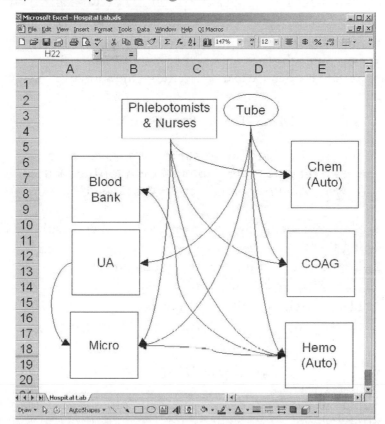

- Rework—fixing stuff that's broken—is one of the more insidious forms of non-value-added work. The customer wants you to fix it, but he or she really didn't want it to break in the first place.
- Requests for new software or a facility change may spend months in a prioritization queue before being worked (non-value-added).
- An order may sit idle waiting on an approval or material.

On a value-steam map, most of the non-value-added time will be found in one of three places:

- *Arrows*—delay between process steps
- *Rework loops*—fixing errors that should have been prevented (e.g., redoing lab or radiology tests)

- *Scrap processes*—discarding or recycling defective products

To eliminate these non-value-added activities, ask yourself how you can

- Eliminate or reduce delay between steps
- Combine job steps to prevent wasteful delay
- Initiate root-cause teams to remove the source of the rework

Map the Value Stream

Purpose: Evaluate the existing or improved process as a starting point for improvement (YouTube video: www.youtube.com/watch?v=3mcMwlgUFjU).

1. Start by identifying customer needs and end with satisfying them.
2. Use square Post-it Notes to lay out processes.
3. Use arrow Post-its to show delays.
4. Place activities in the correct order.
5. Identify inventory levels carried between each step.

Spaghetti Diagrams

Purpose: To examine the existing flow before redesigning it (YouTube video: www.youtube.com/watch?v=UmLrDjT5g8o).

1. Use square Post-it Notes to lay out a floor plan of machines or processing stations.
2. Draw arrows to show movement of product or service through the floor plan.
3. Assess how many times each processing station is used. Is the highest volume closest to incoming materials or products?
4. Identify ways to redesign the flow to reduce unnecessary movement of people and materials.

Figure 2.6 shows an example from a hospital laboratory. There are five main processing areas: hematology, chemistry, coagulation, urinalysis (UA), and microbiology. Many of these areas have both automated analyzers and manual processes.

Notice that although HEMO has 300 orders a day, it's farther from the pneumatic tube than UA, which only has 48 orders a day. Moving HEMO and Chemistry closer to the tube and UA farther from the tube could reduce unnecessary travel for hundreds of samples.

Once redesigned, the hospital lab saved

- 17 percent of floor space
- 54 percent of travel time
- 7 hours of delay per day

WALKING IS WASTE

Because much of healthcare relies on people, the first rule of Lean in healthcare is that *walking is waste* (or what they say in Lean: *Unnecessary movement of people or materials*).

> **Insight: Less walking = more and better care.**

Unfortunately, some of this unnecessary travel is "built in" by the design of the building or unit. I consulted with an architectural firm that had been asked by a long-time client to design two *lean* rural hospitals. The hospital system also had let this request for proposal (RFP) out to other architecture firms. The architecture group had brought in Lean "experts" who dazzled them with their depth of knowledge but didn't help them figure out how to integrate Lean into the new design. I spent a day with them, using Post-it Notes to come up with a few possible configurations. I even suggested that they use a Post-it Note design presentation with their client to engage the client in the overall design.

I had architecture group think about the customers for the building: inpatients, outpatients, clinicians, building maintenance, operating costs, etc. What's the *voice of the customer*? Inpatients want a window; outpatients don't want to get lost in the hospital; building maintenance wants ease of operation; and finance wants to minimize costs of operation.

Traditional rural design was flat and spread out, when maybe the shortest travel distance is *up*. A vertical X-shaped building with plenty of elevators might give inpatients a window and reduce travel time for clinicians and families alike.

One of the limiting beliefs I discovered is that "the lab can be anywhere." No it can't. I've seen blood samples sit on an ED clerk's counter for 10 minutes before they get sent via pneumatic tube to the lab. What if the lab was on a wall adjoining the ED? I've seen outpatients commute football fields from registration to the lab and then on to imaging. What if they were all close together? I've seen OR designs that have only one toilet that's a city block from the farthest ORs. What were the designers thinking?

The ED is a major doorway to the hospital. The ED can't afford to have patients waiting in exam rooms for imaging or lab work. The ED can't afford to be boarding admitted patients. The bottom line is that an ED can't afford to be on diversion. To minimize movement of patients and lab samples, imaging and the lab have to be next to the ED.

Conducting a Walking Is Waste Study

It's easy to figure out how existing unit designs consume clinicians' time:

1. Have everyone in the unit wear a pedometer for a week, and record their travel distances.
2. Analyze the data. Why are people walking so much? Are they commuting for supplies? Medical records?
3. What needs to be rearranged to minimize travel? In a nursing unit, maybe it's a wall that blocks movement or a supply closet or lack of supplies in the patient's room.
4. Change it! Prototype the change on a limited, low-cost basis. If it works, expand it. If it's not quite right, tweak it.

When Japanese counselors come into a manufacturing company, they immediately start moving machines and materials around to streamline flow. You can too. Just do it. Stop putting up with the high cost of unnecessary movement. Stop putting up with workarounds. Change the system to better serve patients and clinicians.

Patient Flow in the ICU

Cincinnati Children's Hospital Medical Center (CCHMC) found that patient flow through the 25-bed pediatric ICU caused "gridlock and bottlenecks." Symptoms of the problem included

- Diverting pediatric patients to the cardiac ICU

- Boarding pediatric patients in the post-anesthesia-care unit (PACU) or ED
- Delaying or canceling cases (e.g., elective surgeries)

In 2006, CCHMC developed a model of patient flow through the pediatric ICU. The hospital found that variation in demand for healthcare services can be random (e.g., from the ED) or nonrandom (e.g., elective surgeries). Elective surgeries varied from zero to 10 a day. Of these, two-thirds had an LOS of 1.27 days, 28 percent stayed almost 4 days, and 11 percent stayed almost 10 days (Figure 2.7).

In 2007, CCHMC then implemented changes to "smooth" patient flow. These changes included (1) limiting elective surgeries to five per day (smaller batch size), (2) limiting complex elective airway reconstructions (longest LOS) to three per day (hejunka—load leveling), (3) scheduling postoperative ICU beds, and (4) instituting a "morning huddle" to anticipate and confirm demand for the ICU. As a result, diversions, boarding, and cancellations became rare events, and the daily census grew from 17 in 2006 to 21 in 2008. *Smoothing patient flow allowed for an increase in capacity without adding beds!* In 2009, CCHMC expanded to a 35-bed ICU and increased elective surgeries to six per day.

Hint: Getting faster and better means more business!

FIGURE 2.7

ICU length-of-stay Pareto chart.

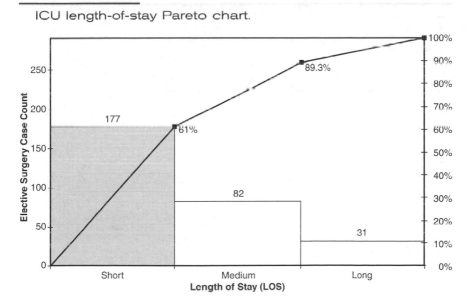

To sustain the improvement, CCHMC monitors a few measures monthly:

- Elective surgeries requiring an ICU bed (daily)
- Number of elective surgery cases in the ICU (average)
- Patients boarded in the PACU or diverted to cardiac ICU (Figure 2.8)
- Elective surgery cancellations owing to ICU bed capacity (Figure 2.9)

LEAN PRINCIPLE–LOAD LEVELING

The CCHMC case study demonstrates one of the Lean principles, namely, load leveling. Leveling the elective surgery load coming out of the OR helps to level the load not only on the ICU but also on the PACU and ED as well.

LEAN PRINCIPLE–MINIMIZE INVENTORY

In 2008, Seattle Children's Hospital had an inventory problem in the ICU. Supplies were so unreliable that nurses started stockpiling

FIGURE 2.8

Days between PACU overnight stays control chart.

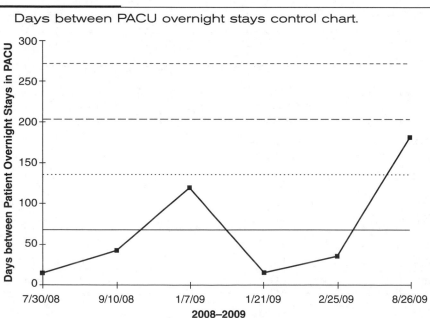

FIGURE 2.9

Days between delayed or canceled surgeries control chart.

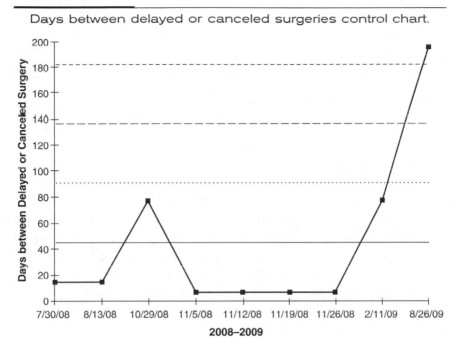

catheters and surgical dressings in patient rooms and offices. This led to greater shortages. Then the hospital went Lean and instituted *kanban* (i.e., inventory system). In a fully stocked rack of color-coded supplies, there are two bins for each supply. When one bin is empty, it goes to central supply for replenishment, where the bar code on the bin is scanned to generate new orders (mistake-proofing). Improvement projects such as this one helped Seattle Children's Hospital to cut costs per patient by 3.7 percent for $23 million in savings. While patient demand has grown, improvement projects such as this one have helped the hospital to avoid $180 million in capital improvement projects by using existing facilities more efficiently, says Patrick Hagan, president.

THEORY OF CONSTRAINTS (TOC)

The *theory of constraints* (TOC) emerged from Goldratt and Cox's book, *The Goal* (1984, North River Press). TOC argues that it makes no sense to run everything in your business flat out if there is a bottleneck or constraint that governs production.

Liebig's *law of the minimum* states that growth is controlled not by the total of resources available but by the scarcest resource—the constraint.

TOC Process

The theory of constraints consists of five steps:

1. *Identify the constraint.* All businesses are chains of systems. The constraint is the poorest-performing link in the chain.
2. *Exploit the constraint.* Optimize its performance. Break "dummy" and "policy" constraints.
3. *Pace every other process to the speed of the constraint.* Reduce overproduction, work in process, inventory, etc.
4. *Invest in the constraint.* If production is still too low, invest in reducing the constraint (e.g., double the number of people or machines performing the constraint process).
5. *Repeat until you get the desired performance.* There may be a new constraint.

Constraints

There are four types of constraints:

- *Resource constraints (people, machines, materials).* In a manufacturing company, one machine may be able to produce only 10 widgets per hour, whereas the other machines can do 100. In a hospital, the number of beds might be a constraint, or the number of physicians or nurses in the ED, or the turnaround time of lab results.
- *Market constraints.* Demand is greater than production capacity (e.g., recession).
- *Policy constraints.* "We've always done it this way, but no one know why." Some human resources departments have a policy that every job opening must be opened to existing employees for a period of time (e.g., five days). The majority of job openings are for entry-level positions that no existing employee wants. Eliminating this policy for entry-level jobs accelerates hiring of entry-level personnel such as bank tellers or call-center representatives.

- *Dummy constraints.* Constraints caused by low-cost, easily expandable resources (e.g., phone lines, faxes, printers, PC software, etc.).

Exploiting the Constraint

There are simple ways of exploiting and breaking the constraint:

- *Resource constraints* are rarely used at 100 percent of capacity. How can you increase utilization to 100 percent? How can you reduce the non-value-added aspects (e.g., wait time) of the resource. Toyota reduced changeover time from an hour to three minutes or less, increasing the use of all machinery. In a hospital, preventing OR cancellations by ensuring that every patient has everything needed to begin surgery leads to greater throughput.

  What bottleneck in your hospital could be optimized to run at full capacity?
- *Market constraints.* It's hard to change the market, but it is possible to discover new applications for existing products that will create new markets.
- *Policy constraints.* Changing call-center measures from length of call (minutes) to first-call yield (one call does it all; no repeat calls) can increase customer satisfaction and reduce the total call volume.

  What archaic policies could be eliminated or rewritten to optimize throughput and patient satisfaction?
- *Dummy constraints.* In a hospital ICU, adding a cleaning person (low-cost relative to RNs) can accelerate the availability of ICU beds. Moving an order printer closer to the CT scanners can speed up exams. Giving office workers their own multifunction printer/fax/scanner can eliminate unnecessary commuting to a centralized printer and eliminate rework caused by people taking more than their own printout.

We often find that hospitals restrict access to the QI Macros to just their improvement specialists. Unfortunately, this shifts their focus from improvement to drawing charts for everyone else. Software is a dummy constraint. Give QI Macros and Excel to every department, and teach them to draw their own charts, thereby freeing up the improvement specialists to focus on improvement.

GET THE IDEA?

A little analysis will quickly reveal the constraints in any system. Then there are some pretty simple and inexpensive solutions to constraints: Eliminate silly, outdated policies; spend a little to eliminate dummy constraints, which will free up more expensive constraints; optimize the use of key resources; and stop overusing non-key resources (overproduction).

If most hospitals operate at a fraction of their capability, what could you accomplish by eliminating unnecessary delays and constraints?

CHAPTER 3

Simple Steps to a Better Hospital

The Institute of Medicine's *To Err Is Human* (National Academy Press, 2000) called for a 50 percent reduction in medication errors, but in 2009, Dr. David Bates said, "With respect to the 50 percent reduction, the truth is that we don't really know, because we don't have good metrics for sorting out how common medical errors are in most institutions." Atul Gawande reported in *Complications* (Holt, 2003) that autopsies turned up major misdiagnoses as the cause of death a shocking 40 percent of the time, and rates have not improved since 1938.

> **Every hospital seems to have the same issues:
> preventable adverse events that will no longer be paid
> by Medicare and other insurers.**

This shows up in many ways:

- Catheter-associated urinary tract infections (UTIs)
- Bloodstream infections (BSIs)
- Surgical-site infections (SSIs)
- Pressure ulcers
- Surgical errors: retained foreign objects, surgical infections, and wrong-site and wrong-patient surgeries
- Blood incompatibility
- Ventilator-acquired pneumonia (VAP)
- Patient falls

A recent RAND study found that only one of every two patients will receive care that meets generally accepted standards: 30 percent of stroke patients, 45 percent of asthma patients, and 60 percent of pneumonia patients. In 2009, the Centers for Disease Control and Prevention (CDC) estimated that 1.7 million health-care-associated infections (HAIs) resulted in 99,000 deaths (271 per day—this is equivalent to a 747 crashing every day) and an additional $35.7 billion to $45 billion in unnecessary costs per year. Between 2004 and 2006, 238,337 preventable deaths occurred involving the Medicare population. In addition, 4.5 patients per 100 admissions will acquire an infection while in the hospital. Types and number of HAIs and estimated costs per patient are as follows: SSIs—290,485 at $34,670; BSIs—92,001 at $29,156; VAP—52,543 at $28,508; UTIs—449,334 at $1,007; and *Clostridium difficile*–associated disease (CDI)—178,000 at $9,124.

> **Question: What one element is critical to both improved outcomes and patient satisfaction?**
>
> **Answer: Reducing defects (i.e., medical mistakes) from the current 30,000 patients per million (PPM) to 3 PPM.**

A BETTER EMERGENCY DEPARTMENT (ED) IN FIVE DAYS

Studies have shown that between 2 and 8 percent of patients with heart attacks who are seen in emergency rooms are mistakenly discharged, and a quarter of these people die or suffer a complete cardiac arrest.

—ATUL GAWANDE

> **Root cause: Misread electrocardiograms (ECGs).**

In *Complications*, Gawande identifies one Swedish study that trained a computer to read ECGs. It outperformed a leading cardiologist by 20 percent.

A BETTER CLINICAL STAFF IN FIVE DAYS

Estimates are that, at any given time, 3 to 5 percent of the practicing physicians are actually unfit to see patients.

—ATUL GAWANDE

Much like a rapid response team (RRT) uses indicators to prevent a code, Dr. Kent Neff identified what he calls "behavioral sentinel events that signal the need for an intervention": persistent anger, abusive behavior, bizarre or erratic behavior, and violating professional boundaries. Another measure is the number of lawsuits or complaints against a physician.

A BETTER OPERATING ROOM (OR) IN FIVE DAYS

> We've celebrated cowboys, but what we need is more pit crews.
>
> —ATUL GAWANDE

There are 50 million surgeries each year in the United States, with 150,000 deaths. Estimates for complication rates range from 3 to 17 percent. Worldwide, there are 230 million surgeries leaving 7 million patients disabled and 1 million dead. Research has shown that half of all complications and deaths are avoidable. Surgery has *four big killers* worldwide:

- Infection
- Bleeding
- Unsafe anesthesia
- *The unexpected*

Atul Gawande, a surgeon at Brigham and Women's Hospital in Boston, authored *The Checklist Manifesto*, a book about using surgical checklists to reduce operation times, infections, and deaths by more than a third. Gawande participated on the World Health Organization's team to develop a surgical safety checklist, including simple things such as having everyone on the surgical team introduce themselves by their first name. "Giving people a chance to say something at the start seemed to activate their sense of participation and responsibility and their willingness to speak up." Operating team members' sense of functioning well as a team jumped from 68 to 92 percent.

A Better OR in Two Minutes

The World Health Organization's (WHO's) Surgical Safety Checklist and implementation manual are available at www.who.int/

patientsafety/safesurgery/ss_checklist/en/index.html. Use of the checklist in eight trial hospitals delivered the following results:

- A 36 percent reduction in major complications
- A 47 percent reduction in deaths
- A 50 percent reduction in infections
- A 25 percent reduction in returns to the OR to fix something missed

At Kaiser hospitals in Southern California using the checklist, the rate of OR nurse turnover dropped from 23 to 7 percent. Employee satisfaction rose by 19%. And staff rating of the "teamwork climate" rose from good to outstanding.

By the end of 2009, 10 percent of U.S. hospitals had implemented the checklist. It takes doctors an estimated 17 *years* to adopt a new treatment.

Gawande says:

In the spring of 2007, I began using [the surgical checklist] in my own operations. . . . did I think the checklist would make much of a difference in my cases? No. In *my* cases? Please. To my chagrin, however, I have yet to get through a week in surgery without the checklist's leading us to catch something we would have missed.

Want to improve surgical outcomes? Use the WHO's surgical checklist and introduce everyone before surgery starts (www.who.int/patientsafety/safesurgery/en/. Checklists can be good medicine for doctors, nurses, and patients.

Retained Foreign Objects

Every 120 minutes, a retained foreign object (RFO) occurs in U.S. surgeries. Retained foreign objects (i.e., surgical left-ins) occur in one out of a thousand (1,000 PPM) abdominal operations, resulting in significant adverse outcomes. In 2005, the Mayo Clinic Rochester averaged one RFO every 16 days. By changing the process for counting and tracking surgical supplies and instruments, the clinic was able to extend time between RFOs to 69 days (Figure 3.1).

With over 100 unique items employed in surgery, which item was left in most commonly? Sponges (Figure 3.2). Changing the

FIGURE 3.1

Days between RFOs chart.

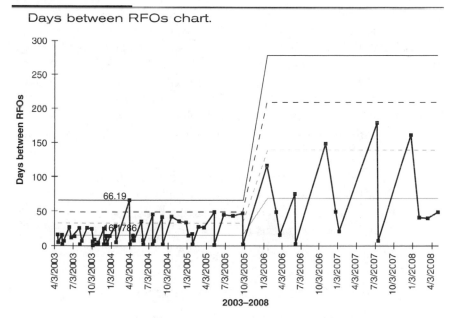

FIGURE 3.2

Retained foreign objects Pareto chart.

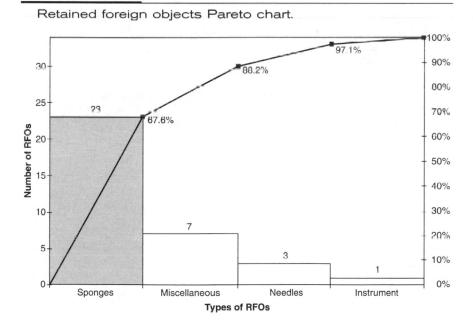

sponge counting process to require an *exact* match of sponges in and out resulted in far fewer sponge RFOs.

Instead of counting manually, why not use technology? ClearCount Medical Solutions (www.medgadget.com/archives/2010/01/markets_first_rfid_surgical_sponge_tracking_system .html) developed Food and Drug Administration (FDA)–approved sponges fitted with a radiofrequency identification (RFID) chip that is smaller than a penny. A handheld wand detects commonly used surgical sponges. Here's what ClearCount identifies as the benefits of using RFID sponges:

- Passive: The nonemitting tag contains no battery.
- Small: The RFID tag is the size of a penny.
- No line of sight is required to detect the sponges.
- The detector can read multiple sponges simultaneously.
- The detector can't count the same sponge twice.

Surgical-Site Infections

Henry Ford Hospital in Detroit initiated tight glycemic control in 2003, reducing SSIs, especially in bariatric patients. (Figure 3.3 note: numbers approximated from article.)

If you go to www.qimacros.com/hospitalbook.html, you can download the QI Macros Lean Six Sigma Software 90-day trial. Click on the "Fill-in-the-Blank SPC Charts" to explore the *g* chart template for Excel. Use the Pareto chart macro to draw Pareto charts.

Wrong-Site or Wrong-Patient Surgery

The Joint Commission reports that there are four to six wrong-site surgeries *per day* (www.centerfortransforminghealthcare.org). Several states require hospitals to report adverse events: Connecticut (www.ct.gov/dph/lib/dph/hisr/hcqsar/healthcare/pdf/adverseeventreportoct2009.pdf), Minnesota (www.health .state.mn.us/patientsafety/publications/index.html), New Jersey (www.state.nj.us/health/ps/documents/ps_initiative_report07. pdf), New York, and Pennsylvania (www.patientsafetyauthority .org). From 2004 to 2009, Pennsylvania wrong-site surgeries averaged 15.73 per quarter. Using a control chart (Figure 3.4) would suggest a process shift in fourth quarter 2008, *when Medicare stopped*

FIGURE 3.3

Surgical-site infections.

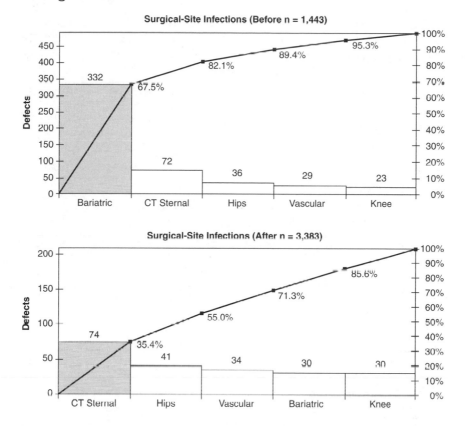

Surgical-Site Infections (Before n = 1,443)

Surgical-Site Infections (After n = 3,383)

paying for treating these mistakes, but we need three more data points to confirm the trend.

The most common type of wrong-site error? Wrong-site anesthesia (Figure 3.5), 29 percent on average.

Possible Countermeasure:
Stop Paying for Medical Mistakes

As of October 2008, U.S. hospitals no longer received Medicare reimbursement for healthcare-associated infections: catheter-associated urinary tract infections (UTIs), central venous catheter–related bloodstream infections (BSIs), and ventilator-associated pneumonia (VAP). In 2001, the Joint Commission analyzed 126 wrong-site or wrong-patient surgeries (Figure 3.6). Most

FIGURE 3.4

Pennsylvania wrong-site surgeries Control chart.

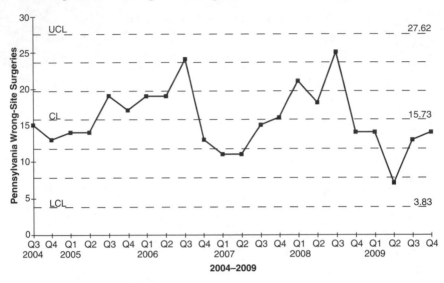

FIGURE 3.5

Pennsylvania wrong-site anesthesia Run chart.

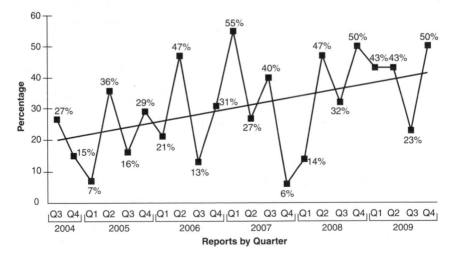

FIGURE 3.6

Type of wrong-site surgery Pareto chart.

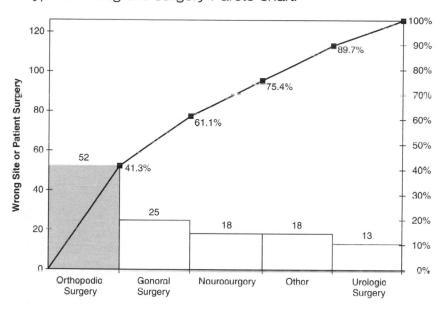

involved orthopedic surgeons and the wrong body part or site (Figure 3.7). This led to the creation of a universal protocol to help prevent these kinds of mistakes (www.jointcommission.org/PatientSafety/UniversalProtocol/).

Another study in 2007 found that "the number of sentinel events reported to the Joint Commission has not changed significantly, despite the required use of the Universal Protocol. Wrong-site surgery continues to occur regularly, especially wrong-side surgery, even with formal site verification." (*Author's note*: if a countermeasure doesn't change performance, it's not a solution.) In one state, over 30 months, there were 427 reported incidents, and 83 patients had incorrect procedures done to completion. Thirty-one formal time-out processes were unsuccessful in preventing wrong surgery. The most common type of incident? Wrong-side surgery (Figure 3.8). Who is most likely to catch the error? Patients and nurses.

Most common root cause: The actions of the surgeon in the OR (92 reports).

Second: Failure of the time-out process (59 reports).

FIGURE 3.7

Wrong-site or wrong-patient surgery Pareto chart.

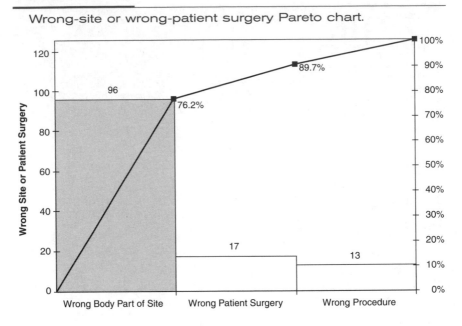

FIGURE 3.8

Wrong-site location Pareto chart.

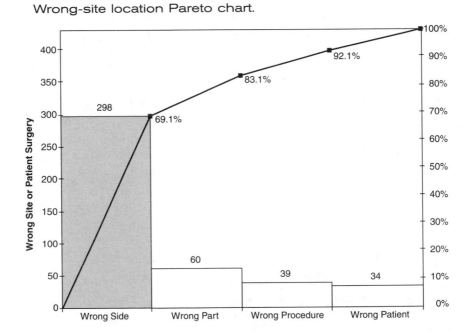

Either of these causes may be a function of confirmation bias (the psychological tendency to confirm an impression despite the facts). Another common thread in wrong-site surgeries: symmetrical body parts such as left/right arm, leg, knee, chest, etc. and positioning of the patient.

In 2005 in Florida, there were 31 wrong-site operations, 5 wrong-patient surgeries, and 86 instances where the wrong procedure was done, according to Dr. Allen Livingstone (Miami, FL).

> **Countermeasure:** The longer the patient is awake before surgery and the greater the involvement of the surgeon and anesthesiologist in preoperative preparation, the greater is the chance of preventing wrong-site or wrong-patient surgeries. Time-outs and the universal protocol don't seem to work that well. What would work better?

A BETTER PHARMACY

With over 400,000 adverse drug events per year costing an estimated $3.5 billion, getting medications right is a big opportunity. There are many types of medication errors: wrong drug, wrong dose, wrong timing, and wrong route, interaction, or patient. At one hospital, medication orders were causing problems. The error rate was 3,300 per 1 million orders.

The most common type of error? Order not received (Figure 3.9). Second runner-up? Wrong frequency of dose. These two accounted for almost half of all order errors.

Most orders were faxed, and fax-line congestion prevented orders from being received. Nurses sometimes missed changes in frequency or dosage.

After implementing a computerized order-entry system and other procedural changes, order errors fell from 3,300 to 1,400 per million, a 55% reduction with an estimated cost savings of $1.2 million per year.

The 4-50 Rule in Medication Errors

As you might expect, some medications are more dangerous than others. *High-alert medications*, including insulin, anticoagulants, narcotics, and sedatives, should trigger a heightened focus on the opportunity for a medication error according to the Institute for

FIGURE 3.9

Pharmacy errors Pareto chart.

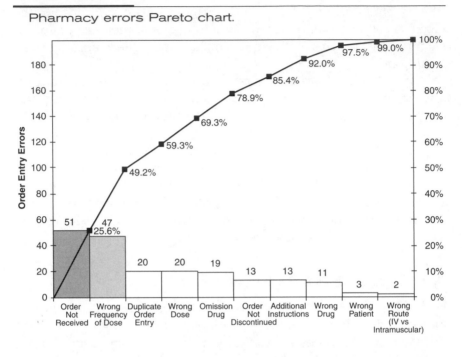

Healthcare Improvement (IHI). From 2006 to 2008, 537 hospitals reported 443,683 medication errors; 32,546 were related to high-alert medications (Figure 3.10). The most frequent error and most frequent cause of harm? Insulin (Figure 3.11).

Where did the errors originate? Dispensing, administering, and transcribing (Figure 3.12). Most common type of error? Omissions, dosage (e.g., 5.0 misread as 50), and wrong drug (Figure 3.13). This is how Dennis Quaid's newborn twins ended up getting 1,000 times the amount of blood thinner prescribed.

Shouldn't there be a way to mistake-proof this process? Some hospitals are adopting computerized physician order entry (CPOE).

BETTER ORDER ACCURACY IN FIVE DAYS

In 2006, Memorial Hermann Baptist Hospital (MHBH) in Beaumont, Texas, a 250-bed acute-care hospital with a 17-bed pediatric unit, faced a disturbing challenge. The city's pediatricians were steering their patients away from the hospital's emergency

FIGURE 3.10

High-alert medication errors Pareto chart.

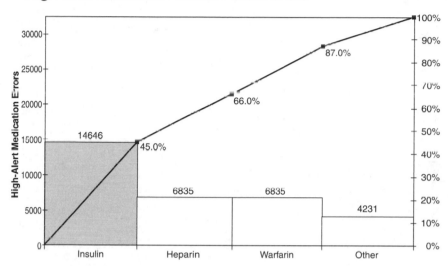

FIGURE 3.11

Harm from high-alert medication errors Pareto chart.

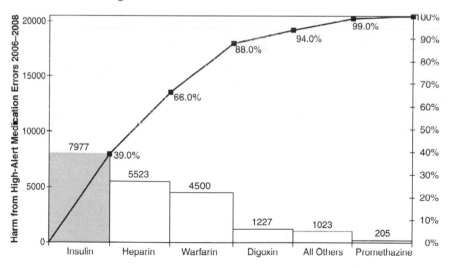

FIGURE 3.12

Origin of medication errors Pareto chart.

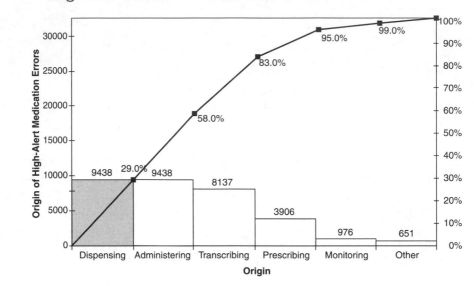

FIGURE 3.13

Type of medication errors Pareto chart.

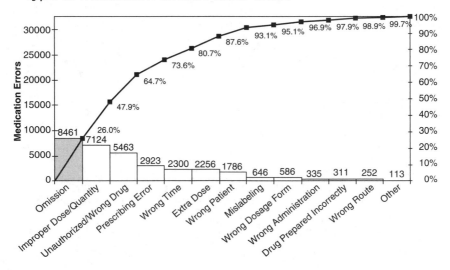

department (ED). In as many as half of cases, the ED's admission orders conflicted with the pediatrician's orders. These conflicts created a crisis of confidence in the pediatric community.

Define

MHBH defined the "defect" in this case as the care provided or ordered by ED physicians differing from or conflicting with what the patient's pediatrician would have found acceptable. The "defects" fell into a few categories: inappropriate medication, dosing, fluid type, fluid dosing, missing diagnostics, and other. The goal was to reduce these conflicts by 75 percent in six months.

Measure

With almost 9,000 pediatric visits to the ED yearly, it took some analysis to narrow the focus. With some digging, it became apparent that *admitted children under the age of 6* represented most of the conflicts. (This could have been shown using a Pareto chart.)

Analysis

The 149 charts reviewed contained 77 defects (a 50 percent error rate). Recognizing that there are many opportunities for error in an admission, the defects per million opportunities (DPMO) was calculated at 80,800, or about a 7 percent error rate (7 children out of 100 received incorrect admission orders). Further analysis found that the errors were in the admission orders but never delivered to the patient. The pediatric nurse would catch them and call the pediatrician for clarification.

Root cause: The admission order, not patient care.

Improve

ED physicians were not pleased that their care was being critiqued, but with some cajoling, they agreed that a possible countermeasure would be a specific order set for pediatric admissions. The pediatric nursing director developed a draft of the orders. The orders were reviewed, revised, and approved by pediatricians in the city.

Within three weeks of implementation, DPMO fell from 80,800 to 15,238, and within three months, it fell to 10,452 (Figure 3.14)—an improvement of 86 percent!

Control

To sustain the improvement, MHBH monitored order-set usage and conducted random chart reviews on a quarterly basis, all of which were reported at ED staff meetings.

BETTER MEDICAL IMAGING IN FIVE DAYS

Peer pressure can be a powerful incentive. One analysis showed wild variation in the use of computed tomographic (CT) scans and magnetic resonance imaging (MRI) in one medical group. After presenting the data, radiology test use fell by 15 percent the first year. Continuous monitoring has held the rate constant.

A BETTER LAB IN FIVE DAYS

As many as two-thirds of lab errors occur in the order and labeling process, before testing begins. In 2003, North Shore Long Island

FIGURE 3.14

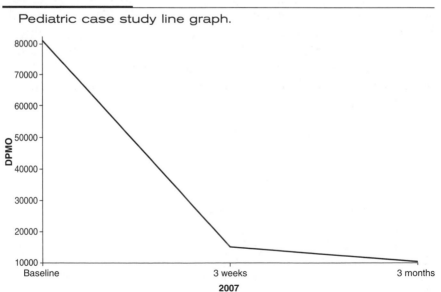

Pediatric case study line graph.

Jewish Health System set out to use Six Sigma to reduce its errors. The hospital found that 5 of 100 samples were inaccurate or incomplete. The team analyzed 5,667 laboratory requisitions and identified 285 errors. The most common? Social Security Number errors in skilled nursing facilities (Figure 3.15).

Root Cause: Skilled nursing facilities used addressographs instead of available bar-code labels for sample identification.

Countermeasure: Use bar-code labels.

The North Shore Long Island Jewish Health System also colorcoded samples and parts of the lab to ensure that samples were delivered to the correct location for processing, saving additional time and reducing errors.

Results

DPMO fell from 7,210 to 1,387. Staff productivity rose from 20 to 23 requests per hour to handle additional volume (Figure 3.16). Combined improvements resulted in $339,000 in increased revenue and cost reduction.

FIGURE 3.15

Lab order errors Pareto chart.

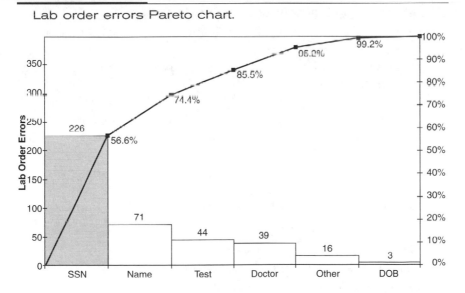

FIGURE 3.16

Lab requisitions per hour control chart.

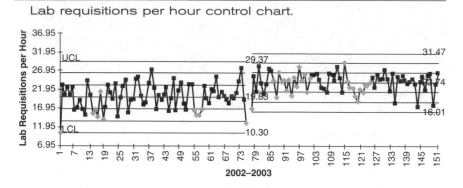

2002–2003

A BETTER NURSING UNIT IN FIVE DAYS

Patient falls can lead to significant morbidity and mortality. The estimated cost to treat serious falls-related injuries ranges from $15,000 to $30,000 per fall. In 2009, Connecticut reported that death or injury caused by patient falls was the most common reported adverse event (Figure 3.17).

Where do patients fall? According to New Jersey statistics, the patient's room (Figure 3.18).

Countermeasures to Prevent Falls

- Formalize falls risk assessment for each patient.
- Create a checklist of medications known to increase a patient's risk of falling.
- Have the pharmacy color-code medications known to increase falls risk.
- Track medication-related falls, and add them to the list.
- Use nursing whiteboards to identify high-risk patients.
- Use hourly rounds to check on high-risk patients (e.g., toilet needs, etc.).
- Use color-coded clips on wheelchairs and stretchers when transporting high-risk patients.
- Provide nurses and physicians with pocket guide to falls prevention.

These countermeasures use simple Lean Six Sigma principles such as (1) make it visual (i.e., color-coded) and (2) use checklists for risk assessment and medications.

FIGURE 3.17

Connecticut adverse events Pareto chart.

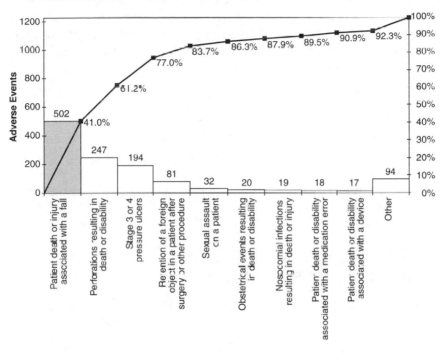

FIGURE 3.18

New Jersey adverse event patient falls Pareto chart.

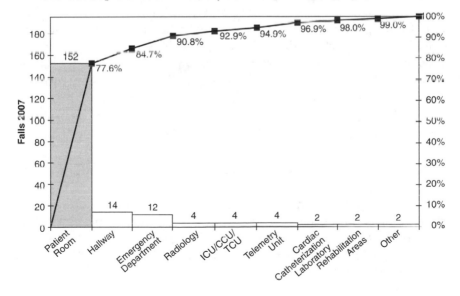

BETTER DIAGNOSES

According to one *BusinessWeek* article, 15 out of every 100 patients are misdiagnosed. Patients return to EDs after being discharged because they still have undiagnosed or untreated symptoms. This is known as *rework*! It also leads to billing problems because of duplicate date-of-service issues.

Solution

- Measure repeat ED visits within 24 hours or misdiagnoses.
- Use the "dirty 30" process (Chapter 5) to identify the root causes.
- Identify ways to mistake-proof the diagnosis and treatment of these patients.

A BETTER INTENSIVE-CARE UNIT (ICU) IN FIVE DAYS

The IHI (ihi.org) estimates that preventable physical harm to patients occurs 40,000 times a day in U.S. hospitals. In addition, 90,000 patients a day are admitted to ICUs; survival rate is 86 out of 100. The CDC estimates that 2 million people are affected by SSIs, drug reactions, and bedsores, and 99,000 people die as a result of HAIs.

Central line blood stream infections (CLBSIs) are a serious problem. Of the 5 million lines inserted each year, about 4 percent (*4-50 rule*) become infected within 10 days (200,000+); up to a quarter of these patients will die as a result, with a resulting cost of $3 billion and 30,000 deaths. The other three-quarters will spend an extra week in the ICU. The average cost per infection is $45,000.

SSM Health Care reduced CLBSIs from 6 per 1,000 central line days in 2006 to 0.4 per 1,000 in 2010. In 2004, Allegheny General Hospital reduced central line infections from 10.5 to 1.2 per 1,000 central line days in its ICUs. Mortality from central line infections fell from 19 to 1.

One hospital found that monitoring infections using fresh needle sticks rather than using blood from the intravenous (IV) line provided a better detection method. This hospital also used colored tape to mark IV lines inserted under less than desirable conditions (e.g., in an ambulance, in the ED, etc.). Such lines were changed as soon as the patient got settled in a nursing unit, which reduced infection rates.

In 2001, Dr. Peter Pronovost at Johns Hopkins Hospital came up with a five-item checklist that reduced central line infections from 11 percent to *zero*. In 2003, the infection rates in 77 Michigan hospitals fell by two-thirds in the first three months of checklist use, saving 1,500 lives and $200 million in the first 18 months. Most ICUs cut their infection rates to zero. "The successes have been sustained for several years now—all because of a stupid little checklist," says Atul Gawande. The checklist included simple solutions such as (1) wash your hands with chlorhexidine, (2) clean the patient's skin with chlorhexidine antiseptic, (3) use a sterile drape over the entire patient, (4) wear a mask, cap, sterile gown, and gloves, and (5) put sterile dressing over insertion site. Dr. Pronovost found that physicians skipped at least one step with a third of patients. So why do doctors resist using checklists?

Unlike pilots, doctors don't go down with their planes.
—JOSEPH BRITTO, MD

Dr. Pronovost went on to establish checklists for

- Administering pain medication that reduced untreated pain from 41 to 3 percent
- Ventilation patients to ensure prescribing antacids to prevent stomach ulcers and elevating the bed 30 degrees to stop secretions from entering the windpipe, reducing pneumonia by 25 percent and preventing 21 deaths

Use of checklists reduced ICU length of stay by *half!* Combining these checklists with the WHO's surgical checklist could slash adverse events. Concise checklists for error-prone processes may be the ultimate safety tool for patients and clinicians. Pilots use them to safeguard air travel. What can we learn from their expertise?

Between 2003 and 2006, Allegheny General Hospital reduced central line infections by 95 percent and deaths from CLBSI to zero. In 2006, the hospital went a whole year without a central line infection. In 2010, Steven and Alexandra Cohen Children's Medical Center of New York went a whole year without a central line infection in the pediatric ICU while treating 1,647 patients for 2,574 central line days. When the medical center started the program, the central line infection rate was 4.7 per 1,000 central line days. In pediatric patients, maintenance of the line accounts for 90 percent of the infections. The medical center implemented a "scrub the hub" of the catheter port with a special cleaning solution, frequent tubing changes, and a new protocol for changing the catheter dressing.

One of the most important shifts was away from the hierarchical medical model to an open culture where, "If you see something, say something." Nurses could point out physician errors such as contaminating a glove during line insertion.

The VA Pittsburgh reduced methicillin-resistant *Staphyloccus aureus* (MRSA) infections from 0.97 per 1,000 patient-days in 2002 to only 0.27 in 2004. (*Author's note:* 0.27/1,000 patient-days is equivalent to 270 per 1 million patient-days—or about 5 sigma.)

MISUSE OF ANTIBIOTICS

While infections are a problem, misuse of antibiotics can lead to other problems. Providence Saint Joseph Medical Center (PSJMC) found that nursing units often failed to discontinue antibiotics within 24 hours of surgery end time for up to 1,000 patients per year. Failure to stop antibiotics can lead to adverse reactions and increased medical costs.

PSJMC found that average stop time for antibiotics was 39 hours after surgery. Only 25 percent of cases were compliant with guidelines. And there was no standard process or protocol used in the nursing units. The medical center also found that orthopedic and colon surgeons had the highest noncompliance rates.

Countermeasures

- Revise order sets with support from surgeons.
- Identify applicable patients in the OR.
- Automate discontinuation of antibiotics by the pharmacy at the twenty-fourth hour for applicable patients.
- Add orange stickers to patient charts to visually identify applicable patients.
- Monitor compliance daily.

In a few months, compliance rose from 36 to 90 percent, which generated $35,000 in savings.

BAR CODES BUST MEDICATION ERRORS

Good news: When the VA adopted bar codes for patients and medicines, medication errors plummeted. By bar coding medications and patients and using handheld scanners, clinicians can ensure that the right patient gets the right dosage of the right medication at the right time.

Bad news: An estimated 7,000 people each year die in hospitals from medication errors. One out of every 14,000 transfusions gets the wrong blood, resulting in at least 20 deaths each year. Only about 125 of the nation's 5,000 hospitals use bar codes.

Good news: The FDA required bar codes on all medications starting in February 2004.

Bad news: The national average for wristband inaccuracies in hospitals is 3 percent. (If you get the band wrong, everything else can go wrong too.)

Sadly, safety technology isn't a big diagnostic machine that generates revenue; it's a protective device that reduces the cost of treatment and litigation. The good news is that the technology is out there to make our healthcare safer than ever before. All we have to do is embrace it.

THE PROBLEM ISN'T WHERE YOU THINK IT IS

The most common root cause of adverse events reported in state statistics is *communication.* By this, healthcare professionals usually mean person-to-person communication. If you want to be faster and better, you cannot rely on person-to-person, mind-to-mind communication. It has to be in the medical record or visual. Could patients with a high risk of falls be given a different-colored gown or detachable tag that travels with them? What systems could be put in place, such as a doctor marking a surgical patient's ID to indicate that the patient has received appropriate checks and instruction before surgery? Could a voice recorder carry a patient's status from the ED to a nursing floor? Stop thinking fleeting mind-to-mind communication, and start thinking visual and mechanical systems.

In any industry, including healthcare, managers and employees always think that better training will solve their quality problems. Unfortunately, training doesn't always stick, and turnover drains the skill pool. The only way to prevent errors is to build the prevention into the systems and procedures. This means implementing *standard* procedures, checklists, and measurements to monitor performance. It also means endlessly tuning these procedures, checklists, and measures to improve performance. If you implement a countermeasure and it doesn't reduce errors, then it's not a good countermeasure. Stop doing it, and implement something better.

Key insight: Processes, procedures, and systems cause most medical mistakes and errors, not people.

The goal: Eliminate mistakes and errors by changing processes.

One in five Medicare patients is readmitted within 30 days but hasn't seen a doctor before they return (Figure 3.19). More than 50 percent are readmitted or die within a year, a defect in ongoing care. Most common readmissions: heart failure, pneumonia, chronic obstructive pulmonary disease (COPD), psychosis, and gastrointestinal (GI) problems. Estimated cost in 2004: $17.4 billion. Reducing readmissions is one of the cost-saving initiatives in healthcare reform.

FIGURE 3.19

Medicare rehospitalization percent of patients.

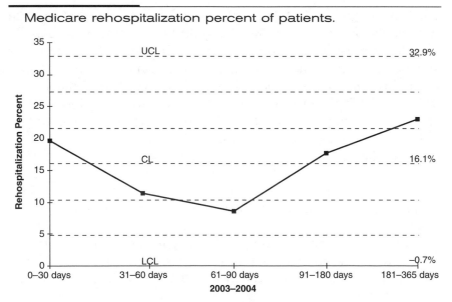

"Hospitals and physicians may need to collaborate to improve promptness and reliability of follow-up care." One Colorado hospital used a *transition coach* for the first 30 days after discharge, reducing readmissions by 20 to 40 percent (www.caretransitions.org).

I PLEDGE ALLEGIANCE TO SCIENCE AND EVIDENCE

At the 2006 IHI annual conference, Don Berwick asked attendees to *pledge allegiance to science and evidence.* It's been over a decade since the Institute of Medicine's *To Err Is Human.* Isn't it time to start capturing every medical error and using that data not to punish the mistake makers but to change systems to prevent the error *forever*?

The Cost of Medical Harm

An estimated 50 patients out of every 100 will suffer some sort of harm during their hospital stay. Some of these "harms" are minor, but many cause temporary or lasting disability or even death.

> **From a Six Sigma perspective, this process is worse than 1 sigma (300,000 PPM).**

With an estimated 37 million hospital admissions a year (and perhaps three times that many ED visits in which the patient is not admitted), *medical harm affects 17 million patients and their families each year.*

The Goal

The IHI's 2006 Five Million Lives campaign hoped to reduce this number by 5 million over two years, or 2.5 million per year. The campaign focused on the top five categories of medical harm (leaving out the minor categories):

A. Temporary injury from care (an estimated 60 percent of the overall total)

B. Temporary injury that requires hospitalization

C. Permanent injury

D. Injury requiring intervention within 1 hour to save the patient's life

E. Death

Campaign Focus

There are six "planks" in this campaign:

1. Prevent pressure ulcers.
2. Prevent MRSA infections ($2.5 billion per year).
3. Prevent high-alert medication errors.
4. Prevent surgical complications.
5. Prevent congestive heart failure complications ($29 billion per year).
6. Get hospital boards on board with the changes required.

There's already plenty of science and evidence that points the way toward solutions that will prevent these types of errors, harm, and injuries. Over 3,000 hospitals have committed to implementing these proven methods. Now comes the hard part: *implementing the change*.

Campaigns

This campaign focuses on hospitals. It doesn't even touch clinics, rural healthcare, doctor's offices, mental health facilities, or most of the other care that occurs in this country. Instead of waiting on the IHI to launch an initiative, I hope that these groups will craft their own campaigns and get started. Find out more at ihi.org.

If you want to know why healthcare is so expensive, the answer may be that there are too many preventable mistakes.

St. Vincent's Medical Center, a 528-bed facility in northeast Florida, reduced pressure ulcers from over 2 to less than 1 per 1,000 patient-days between August 2004 and February 2006. The center's SKIN Program was shared and spread to Ascension Health's other 66 hospitals.

Design Your Own Campaign

One thing I've learned from watching the IHI's 100,000 Lives campaign is that a lot of progress can be made quickly across huge geographic and demographic boundaries by setting clear targeted goals that focus around a shared purpose. Ask any doctor, nurse, or clinician; they all want to serve the patient, even if it means strapping on what Berwick calls the "handcuffs and straightjackets" of rigorous procedures that ensure every patient gets proven therapies (e.g., aspirin at arrival for heart attacks).

What's the overriding purpose in your hospital? What does everyone agree on? How can you craft a campaign to reduce the "harm" (i.e., delay, defects, and deviation) your hospital processes inflict on your patients? How can you craft the campaign so that it will ignite the passion and creativity of your employees and get them to pledge allegiance to science and evidence?

How much progress could you make in the next 24 months?

SIMPLE STEPS TO A BETTER HOSPITAL IN FIVE DAYS

The only realistic hope for substantially improving care delivery is for the old guard to launch a revolution from within. Existing players must redesign themselves.
—RICHARD M. J. BOHMER

How is it possible to get a better hospital in five days or less?

1. *Gather a team that believes it's possible to prevent existing problems* (e.g., an ED team, a nursing unit team, a pharmacy team, etc.). Some people just don't believe it's possible; if so, they won't be useful on the team. Don't load the team with skeptics.
2. *Focus on the improvement.* Gather and graph mistake or error data. Do as much of the analysis as possible beforehand. Key graphs: Control charts of performance over time (usually a c, g, p, u, or XmR control chart) and Pareto charts of mistake categories. One or more teams focus on reducing each "big bar" of the lowest-level Pareto charts.
3. *Have a trained facilitator assist the team in root-cause analysis* (Why? Why? Why? Why? Why?). Have the team identify possible countermeasures to these problems.
4. *Improve.* Implement the countermeasures, and measure the results.
   - Implement process-oriented improvements immediately.

- Implement methods, checklists, and so on to mistake-proof care regardless of the provider.
- Manage more complicated changes (e.g., information technology changes, hardware changes, etc.).

5. *Verify that the countermeasures actually reduce error rates.* (Sometimes they don't.)

6. *Sustain the improvement.* Standardize the improved methods and measures as a permanent way of doing things.

7. *Measure and monitor error rates using control charts to ensure continued peak performance.*

CHAPTER 4

Reducing Defects
with Six Sigma

Using simple charts, I've helped teams save $20 million in just one year. The concept is over 100 years old. It's referenced often in every management book ever written. And it all began with an Italian economist who noticed a simple disparity in incomes. But few people know how to leverage its power in business.

In 1995, I was struggling with the complexities and inherent problems of a huge telecommunications company. Sitting at my desk, I was staring at a vast spreadsheet of facts and figures about postage costs for the company's bills. With a cranky vice president sitting upstairs waiting for my results, I decided to do a simple analysis. There had to be a pattern to the steady increases in postage costs. It wasn't the postal service or ride-along coupons, so it had to be something else.

More and more of the company's bills were being mailed at the 2-ounce rate instead of the 1-ounce rate. Either the postage meters were wrong, or the bills were getting heavier.

On top of this, every month, thousands of bills were returned because of bad addresses. I went through the bins of returned bills looking for any bill over the 1-ounce rate. One by one, I opened them up, looking for the secret to unlock the root cause of increased postage costs.

It took only a few dozen bills to discover the culprit. The company had begun billing for smaller telecommunications companies. Each company got its own page in an already thick envelope. Page by page and company by company, the bill was steadily creeping over the 1-ounce limit. Of course, the product manager who sold the billing service hadn't priced it to cover the increases in postage costs.

Based on my research, a team redesigned the bill to be smaller, lighter, and more readable. In the year after its implementation, postage costs fell by $20 million a year.

What charts did I use to display the problem and garner support for the redesign?

Control charts, Pareto charts, and fishbone diagrams.

I used them again to save $16 million a year in billing adjustments. And again to save $3 million a year in service-order errors. And $5 million a year in denied claims and $20 million in rejected claims for a hospital system.

Isn't it time that you discovered the power of control charts, Pareto charts, and fishbone diagrams to laser-focus your improvement efforts?

INVISIBLE LOW-HANGING FRUIT

When my company first got into the quality-improvement movement in 1990, my Florida Power and Light consultants always spoke about "low-hanging fruit just waiting to be plucked." Two years and tens of thousands of staff hours later, I still hadn't found any low-hanging fruit.

In any company, if there really is low-hanging fruit, it's usually visible everywhere from the emergency department to the management conference room. When it's *that* visible, anyone can pluck it with a little common sense and a bit of trial and error.

This is why in most companies there is no *visible* low-hanging fruit. Somebody has already plucked it! And this is what stops most leaders from even considering the tools of Six Sigma; they can't see any more fruit to be picked.

In company after company and hospital after hospital, though, I have found orchards filled with low-hanging *invisible* fruit. You just can't see it with the naked eye. You can, however, discern it through the magnifying lens of control charts and Pareto charts. *They make the seemingly invisible visible.* They are the microscopes, the MRIs, the ECGs of business diagnosis.

When Louis Pasteur said that there were tiny bugs in the air and water, everyone thought he was crazy because they weren't visible to the naked eye. Everyone thought it was just an "ill wind" that made people sick.

In today's tough economic times, everyone laments about how hard it is. How an "ill wind" has blown through their business, their industry, and their economy. But have they considered using the modern tools of business medicine to root out the infectious agents in their businesses? Have they taken the time to look for the "invisible" low-hanging fruit in their businesses? I doubt it.

Someone sent me an e-mail today that said that even in the poorest-run companies, he'd had no luck finding the low-hanging fruit. But in every company I've ever worked with, I've found millions of dollars just waiting to be retrieved from the caldrons of defects, deviation, and delay. Are you looking for the obvious? Or investigating the invisible?

The low-hanging fruit is always invisible to the naked eye. Turn the magnifying and illuminating tools of Six Sigma on your most difficult operational problems, and stare into the depths of the unknown, the unfamiliar. You'll invariably find bushels of bucks just waiting for a vigilant harvester.

SIX SIGMA'S PROBLEM-SOLVING PROCESS

As you can see, measures, counts, and data about defects and their origins drive Six Sigma's defect-reduction process. Without data about defects or deviation, Six Sigma just won't work. I'm going to suggest that you start with some existing data about defects, mistakes, or errors in some part of the hospital's clinical or operational environment. Most teams get stuck in this *define* and *measure* stage and never get on to the *analyze* and *improve* stage. Start with a real problem about which you have some real data, and you're halfway to success. Then you can leap in to the *analyze*, *improve*, and *control* stages.

DMAIC

While I still think of improvement as following the FISH process— *focus*, *improve*, *sustain*, and *honor*, the Six Sigma problem-solving process uses the acronym DMAIC ("duh-maic"), which stands for

1. *Define* the problem.
2. *Measure* the problem (defects or deviation).

3. *Analyze* the root causes of the problem.

4. *Improve* the process (i.e., implement some countermeasures and verify results).

5. *Control* the process (i.e., measure and monitor to sustain the new level of improvement)

I lump Six Sigma's define and measure steps into *focus*. If you don't laser-focus your improvement efforts using real data about defects or deviation, you aren't doing Six Sigma; you're just doing some version of gut-feel, trial-and-error, knee-jerk problem solving. Or you're trying to retrofit your old way of doing things to look like you're doing Six Sigma. Far too many people start with their pet solution to a problem and try to work their way back to the data that will prove their solution. Too few people start from the data and see where they lead.

GETTING TO LEAN SIX SIGMA

Most successful hospitals that have been around for more than five years get down to about a 1 percent defect rate. I have found from experience that you don't need a lot of exotic tools to rapidly improve from these levels. Companies I've worked with have used these tools to go from 15 percent defects to 3 percent or less in about 6 months and 3 percent to 0.03 percent in about 2 years. Once you get to this level, you'll be ready to use more exotic tools to design your work processes for Lean Six Sigma. Until you get there, though, you may not have the discipline, desire, or rigor needed to use the more advanced tools.

KEY TOOLS FOR DEFECT REDUCTION

There are three key tools in reducing defects, errors, and mistakes:

- *Control charts*—to measure customers' critical-to-quality (CTQ) requirements
- *Pareto charts*—to focus the root-cause analysis
- *Fishbone (Ishikawa) diagrams*—to analyze the root causes of the problems or symptoms

With these three tools you can solve 90 percent of the problems associated with defects, mistakes, errors, or cost—the key problems facing healthcare today.

PROBLEM-SOLVING PROCESS

The Six Sigma problem-solving process (Figure 4.1) also follows the FISH model—*focus*, *improve*, *sustain*, and *honor*. It focuses on identifying problems, determining their root causes, and implementing countermeasures that will reduce or eliminate the waste, rework, and delay caused by these problems.

FIGURE 4.1

Six Sigma problem-solving process.

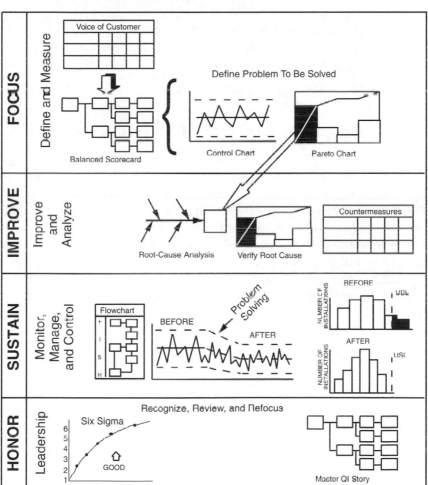

So let's look at how to apply the problem-solving process to achieve Lean Six Sigma improvements in quality and cost. The steps include

1. *Define a problem for improvement* using measurements shown as control chart and Pareto charts to select elements for improvement.
2. *Use the cause-and-effect diagram* to analyze root causes. Then verify and validate the root causes.

Tip: Teams with unvalidated root causes can ruthlessly pursue worthless countermeasures, wasting time and money in the process.

3. *Select countermeasures* to prevent the root causes, and evaluate results from implementing the countermeasures.
4. *Sustain the improvement* using control charts.
5. *Replicate the improvement.*

Critical-to-Quality (CTQ) Measures

There are only two types of defect-related problems—not enough of a good thing or too much of a bad thing, either of which should be measurable and easily depicted with a control chart. Since an increase in the "good" is often a result of decreasing the "bad," measures of the unwanted symptom make the best starting place for improvement. Most hospitals want to report patient *satisfaction* measures, not patient *dissatisfaction* measures, but measures of dissatisfaction tell you where to improve.

Since reducing the unwanted results of a process is often the best place to begin, the area of improvement usually can be stated as reduce defects, mistakes, errors, rework, or cost in a product or service. These are often two sides of the same coin:

An increase in . . .	Is equal to a decrease in ...
Quality	Number defective
	Percent defective
	DPMO—defects per million opportunities
Profitability	Cost of waste, scrap, and rework

Solving problems is usually easiest when you focus on decreasing the "bad" rather than increasing the "good." What are some of the current problems in your work area? Are these problems due to defects, deviation, or cost? Some examples include

- Complaints are defects.
- Missed commitments (e.g., operating room cancellations or start times) are both defects and time problems.
- Waste and misuse of supplies, floor space, computers, networks, or people are cost problems.
- Taking care of a patient injured by a medical mistake, fall, or other preventable error is rework.
- "Never events" are a defect.

How could these be measured and depicted in a control chart to form the basis of an improvement story? Figure 4.2 presents an example of data using vaginal births, costs, and adverse events.

Using Excel's PivotTable function, it's easy to summarize these data by day, doctor, diagnosis, sex, length of stay (LOS), age, etc. Once summarized, these data can be drawn as a control chart of charges or adverse events. We could, for example, compare charges by physician [in QI Macros, sort the data by physician (MD) and procedure date, insert a blank row between physicians, and draw an XmR control chart] (Figure 4.3).

MD1 and MD11 have higher costs and higher variability. MD8 consistently delivers a lower cost with less variability. Is there something we can learn from MD8?

FIGURE 4.2

Maternity data

	A	B	C	D	E	F	G	H	I	J	K
1	DRG	Doctor	DRG	APS DRG	Diagnosis	Age	Sex	LOS	Total Charge	Date	Adverse Event(s)
2	373: Vaginal Delivery	MD11	373	3730	664.01: First-Degree Peri	26	F	2	$ 5,729	10/01/06	--
3	373: Vaginal Delivery	MD11	373	3730	645.11: Post Term Pregnan	18	F	2	$ 9,551	10/02/06	--
4	373: Vaginal Delivery	MD8	373	3730	663.31: Oth Unspec Cord E	37	F	1	$ 6,976	10/02/06	--
5	373: Vaginal Delivery	MD8	373	3730	650: Normal Delivery	19	F	1	$ 4,589	10/03/06	--
6	373: Vaginal Delivery	MD1	373	3730	650: Normal Delivery	28	F	2	$ 11,033	10/04/06	--
7	373: Vaginal Delivery	MD2	373	3730	663.31: Oth Unspec Cord E	27	F	1	$ 7,002	10/04/06	--
8	373: Vaginal Delivery	MD3	373	3730	646.81: Oth Spec Complica	24	F	2	$ 7,190	10/04/06	--
9	373: Vaginal Delivery	MD3	373	3730	645.11: Post Term Pregnan	21	F	2	$ 6,313	10/04/06	--
10	373: Vaginal Delivery	MD5	373	3730	650: Normal Delivery	19	F	1	$ 6,377	10/04/06	--
11	373: Vaginal Delivery	MD11	373	3730	656.61: Excessive Fetal G	22	F	1	$ 7,778	10/05/06	--
12	373: Vaginal Delivery	MD3	373	3730	664.01: First-Degree Peri	19	F	1	$ 6,753	10/05/06	--
13	373: Vaginal Delivery	MD6	373	3730	663.31: Oth Unspec Cord E	22	F	1	$ 8,369	10/05/06	Complication
14	373: Vaginal Delivery	MD6	373	3730	664.11: Second-Degree Per	24	F	2	$ 7,292	10/05/06	--

FIGURE 4.3

MD control chart of cost variation.

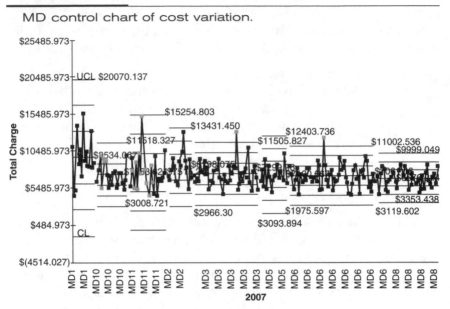

If you go to www.qimacros.com/hospitalbook.html, you can download the QI Macros Lean Six Sigma Software 90-day trial. Use the PivotTable Wizard to create pivot tables. Use the Control Chart Wizard to draw control charts.

Pareto Charts to Focus Improvement

Problem *areas* are usually too big and complex to be solved all at once, but when we whittle them down into small enough pieces, we can fix each one easily and effectively. This step uses the Pareto chart (a bar chart and a cumulative line graph) to identify the most important problem to improve first. Often, two or more Pareto charts are needed to get to a problem specific enough to analyze easily.

Having the control chart of current performance, you'll want to analyze the contributors to the problem. A Pareto chart might take any of the following forms based on the original data:

- *Defects*—Types of defects (e.g., what kind of medicine is most often associated with a medication error?)
- *Deviation*—What's the reason for variation in door-to-doctor or door-to-balloon (angioplasty) time?

- *Cost*—Types of preventable costs, that is, rework or waste (e.g., radiology rework)

What to Look for in Your Data

Most hospitals have lots of data but sometimes have a hard time figuring out what to do with them. I've found that I often use a common strategy for analyzing a company's data. I usually slice and dice an Excel table in the same way:

1. I sometimes have to summarize the defect, error, and mistake data using Excel's PivotTable function.
2. I use Pareto charts to analyze the "total" rows and "total" columns.
3. Then I use Pareto charts to analyze the biggest contributor in each total row or column.

Figure 4.4 presents data about vaginal births, costs, and adverse events. Using Excel's PivotTable function, we can summarize the adverse events into a table and then use the data to create a Pareto chart of adverse events (Figure 4.5).

MD6, MD1, and MD3 account for 72.4 percent of all adverse events. MD1 had 6 out of 15 adverse events, whereas MD8 had only 1 adverse event in 40 deliveries. Fewer adverse events lead to lower costs and less variability. Again, what can we learn from MD8?

What Else? We could evaluate complications by diagnosis using pivot tables (Figure 4.6) and Pareto charts (Figure 4.7).

There are many other ways to look at these data: average LOS by physician or complication. And so on. Get the idea? Pareto

FIGURE 4.4

MD pivot table.

	A	B	C	D	E	F	G	H	I	J
4	Adverse Event(s)	MD1	MD10	MD11	MD2	MD3	MD5	MD6	MD8	Grand Total
5	Complication	3	1		2	5	1	9	1	22
6	Complication, Outlier ($)	1								1
7	Outlier ($)	1		1	1	1				4
8	Readmission				1					1
9	Readmission, Outlier ($)	1								1
10	Grand Total	6	1	1	4	6	1	9	1	29

FIGURE 4.5

Adverse events by MD Pareto chart.

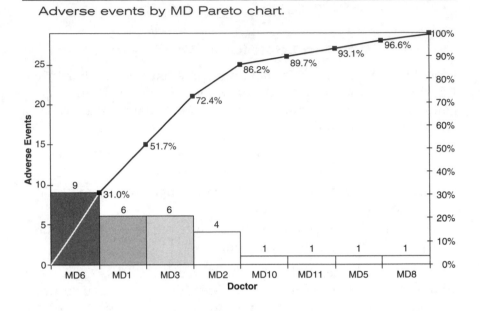

FIGURE 4.6

Adverse events by diagnosis pivot table.

	A	B	C	D	E	F	G
3	Count of Adverse Event(s)	Adverse Event(s)					
4	Diagnosis	Complication	Complication, Outlier ($)	Outlier ($)	Readmission	Readmission, Outlier ($)	Grand Total
5	645.11: Post Term Pregnan	3	1	1			5
6	648.91: Oth Current Condi	2		2			4
7	654.21: Previous Cesarean			1			1
8	656.61: Excessive Fetal G					1	1
9	656.71: Oth Placental Con	1					1
10	660.41: Shoulder (Girdle)	1					1
11	661.21: Oth Unspec Uterin	1					1
12	663.11: Cord Around Neck,	1					1
13	663.31: Oth Unspec Cord E	1					1
14	664.01: First-Degree Peri	1			1		2
15	664.11: Second-Degree Per	3					3
16	664.21: Third-Degree Peri	2					2
17	664.31: Fourth-Degree Per	1					1
18	664.41: Unspec Perineal L	1					1
19	665.41: High Vaginal Lace	2					2
20	665.51: Oth Injury To Pel	2					2
21	Grand Total	22	1	4	1	1	29

FIGURE 4.7

Adverse events by diagnosis Pareto chart.

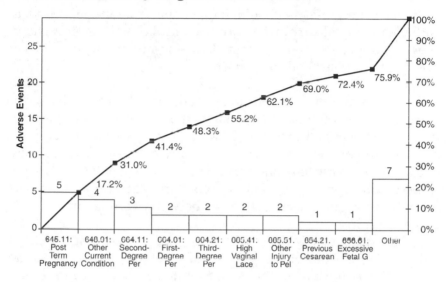

charts are power tools for finding the 4 percent of your business that's causing over 50 percent of the waste, rework, and lost profit.

Eliminating "Never Events"

The National Quality Forum (qualityforum.org) has identified various medical errors as "never events"—they should *never* happen. These include

- *Surgical events* such as wrong-site or wrong-patient surgery, left-in objects, or death
- *Product or device events* such as disabilities arising from contaminated drugs or malfunctioning devices
- *Patient-protection events* such as children discharged to the wrong person, patient disappearance, or patient suicide
- *Care-management events* such as death or disability associated with medication errors, incompatible transfusions, hypoglycemia, pressure ulcers, and labor and delivery
- *Environmental events* such as death or disability from shocks, burns, falls, restraints, or toxic substances

- *Criminal events* such as abductions, assaults, or care provided by someone impersonating a licensed healthcare provider

Since these events should happen infrequently, it's hard to gather enough data to draw a normal control chart. There are, however, a couple of charts that will assist in monitoring "never events": the *g* chart and the *t* chart.

Time-Between Chart—t Chart. The time-between chart, or *t chart*, as you might imagine, measures the time between infrequent (i.e., never) events. Using the QI Macros, it's easy to develop a *t* chart for any of these never events. Simply enter the date and time of each event in the left column, and the template will calculate the elapsed time between events and convert that into a chart. The example in Figure 4.8 shows days between cardiac arrests in a pediatric unit.

The average is 52 days, with an upper control limit (UCL) of 561 and lower control limit (LCL) of 0. Obviously, the hospital would like to move these limits up as much as possible and reduce the variation.

As you can see, some form of improvement took place at the end of 2004. In this case, the hospital implemented family-activat-

FIGURE 4.8

t chart of days between cardiac arrests.

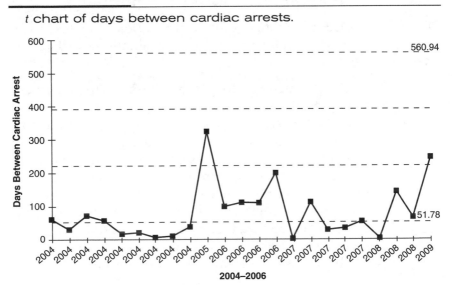

2004–2006

ed rapid-response teams (RRT) in pediatric units. It also looks like the improvement was sustained throughout 2006 but fell back to past performance in 2007. This is not unusual. Most improvements, unless they are woven into the fabric of the hospital, are lost over time. Then the rigor was reinstated in 2008–2009.

Another way to chart these data is with a *g chart*.

Geometric-Mean Chart—g Chart.   The *g chart* can handle the data in two forms:

- Time between never events (just like the *t* chart)
- Number of procedures between never events (e.g., number of surgeries between wrong-site surgeries)

Using the *t* chart data yields a similar chart (Figure 4.9). The QI Macros *g* chart template has the ability to show one process change, in this case, a change between 2004 and 2005. Notice how the average days between rose from 28 to 104, a significant improvement.

Wrong-site and wrong-patient surgeries are just one type of never event—an error that should *never* occur. How can you analyze and prevent never events?

FIGURE 4.9

Days between cardiac events *g* chart.

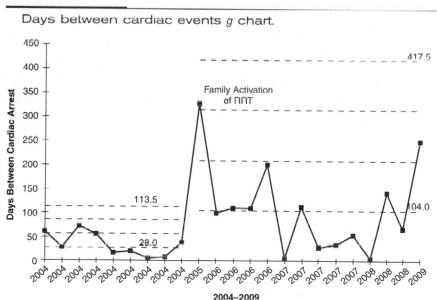

1. Track each never event with g control chart.
2. Use Pareto charts to analyze the most common contributors to the never event.
3. Analyze root causes of "big bar" (i.e., most frequent or most problematic) never events.
4. Implement countermeasures, and verify results.
5. Monitor improvement *forever* using a g chart.

If you go to www.qimacros.com/hospitalbook.html, you can download the QI Macros Lean Six Sigma Software 90-day trial. Click on "Fill-in-the-Blank Lean Tools" to explore the g chart template for Excel.

Suggestions for Improvement

If you spend all your time calculating totals and averages manually, you may miss the point of what you're trying to do, or you may avoid the task completely. Learn to use pivot tables.

If you choose the wrong chart or the wrong data for your chart, you can end up trying to solve problems that don't exist. Learn to choose the right chart for your data.

Use Pivot Tables and the QI Macros Together to Spotlight Your Course of Action

- *Pivot tables.* Use pivot tables to summarize your data in multiple ways.
- *Pareto charts.* Pareto charts are great for showing differences between categories, *not time.* Use them to identify the "vital few," not the "trivial many." Focus on the 4 percent of your defects, mistakes, errors, or categories that provide opportunities for significant improvement.
- *Control charts.* Use control charts to graph your defect or error data over time to monitor performance.
- *Stair-step control charts.* Use control charts with stair steps to show differences in variation and central tendency across multiple sources (e.g., physicians).

Checksheets.   What do you do if you don't have any data to narrow your focus? I find that the best choice is to use a *checksheet* (Figure 4.10). A checksheet can be as simple as a map of the hospi-

FIGURE 4.10

Checksheet for data.

	A	B	C	D	E	F	G	H	
1					Week				
2	Defect/ Problem/ Symptom	M	Tu	W	Th	F	Sa	Total	
3	Delay							0	
4	Missed Commitments	|||||					6	
5	Defects							0	
6	Errors							0	
7	Repeat Repairs							0	
8								0	
9								0	
10								0	
11								0	
12							✛	0	
13	Total	6	0	0	0	0	0	6	
14	http://www.qimacros.com/qiwizard/checksheet.html								

tal or nursing unit with hash marks on it showing where a fall occurred in a nursing unit or a matrix of stroke tallies. Have your "doers" make a mark every time they encounter a certain type of problem. Checksheets can be your friend when you don't have enough data.

Root-Cause Analysis

For every thousand hacking at the leaves of evil, there is one striking at the root.

—HENRY DAVID THOREAU

The *Ishikawa*, *cause-effect*, or *fishbone diagram* helps you to work backward from the problem to diagnose root causes. For those unfamiliar with root-cause analysis, learning to use the fishbone diagram can be frustrating, but once learned, it helps to prevent knee-jerk reactions and symptom patching. There are two main types of fishbone diagrams. One is a customized version of the generic—people, process, machines, materials, measurement, and environment (Figure 4.11). The other is a step-by-step process fishbone that begins with the first step and works backward (Figure 4.12) because errors early in the process often cause the biggest effects. Another tool that's sometimes used to identify causes and effects is the cause-effect matrix (Figure 4.13).

FIGURE 4.11

Traditional fishbone diagram.

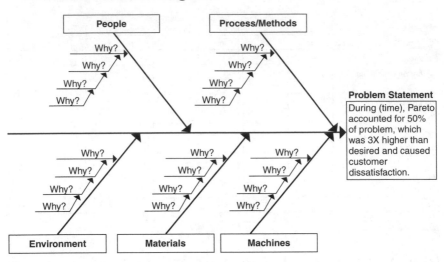

FIGURE 4.12

Step-by-step fishbone diagram.

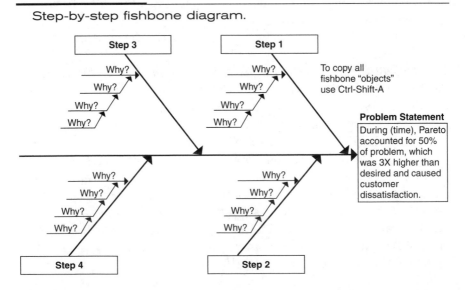

FIGURE 4.13

Cause-effect matrix.

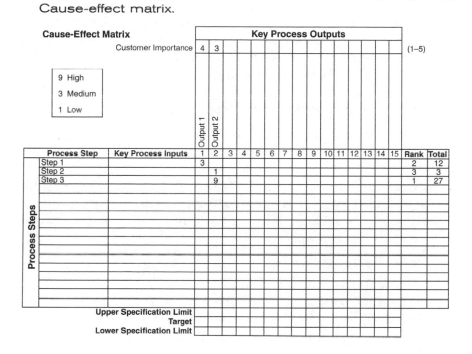

Not only is the fishbone diagram useful for identifying the root cause of recurring problems (common-cause variation), but it also can be extremely useful when stabilizing a process. Special causes of variation (e.g., never events, adverse events, etc.) result in unstable processes as well.

Major contributors to this problem can be identified and root causes determined. When collected over time, these special cause analyses will give you the data to cost-justify the improvements necessary to prevent them.

Fishbone Tar Pits. There are two main tar pits that teams fall into—whalebone diagrams and circular logic.

- A *whalebone diagram* (dozens or hundreds of bones) means that the problem wasn't laser-focused in the first step of analysis. Go back and develop one more Pareto chart at a lower level of detail.
- *Circular logic* (*C* causes *B* causes *A* causes *C* again) invariably means that the logic wasn't checked as it was devel-

oped. Remind participants to ask "Why?" up to five times as you develop each "bone." Then check your logic each time you add a "bone" by working up the chain saying "*B* causes *A*." If the why of *A* is *B*, but *B* does not cause *A*, then the logic is faulty. Remind team members to verify their root causes before proceeding.

Identify and Verify the Root Causes

Take away the cause, and the effect ceases.

—CERVANTES

Like weeds, all problems have various root causes. Remove the roots, and like magic, the weeds disappear. Root-cause analysis simply involves asking "Why?" five times until the root cause reveals itself. For example:

Why did the patient fall? Because the bathroom is too far from the bed.

Why is the bathroom too far from the bed? Because the room was designed to hold two beds but later was converted to a single room, but the bed is placed close to the window, not the bathroom.

Why was the bed placed close to the window and not the bathroom? Because patients want to see outside.

Why do patients want to see outside? (*Hint:* When you get to a silly answer, stop.)

Possible countermeasure: Since the room was designed for two beds, could the bed be moved closer to the bathroom and still allow the patient to see outside?

Defining Countermeasures

Big problems are rarely solved with commensurately big solutions. Instead, they are most often solved by a sequence of small solutions. Small changes can snowball to big changes.

—CHIP AND DAN HEATH

Purpose: Identify the countermeasures required to reduce or eliminate the root causes.

Like good weed prevention, a countermeasure prevents problems from ever taking root in a process. A good countermeasure not only eliminates the root cause but also prevents other weeds from growing.

Never consider a proposed countermeasure to be a solution until you've verified that it worked to reduce falls, medication errors, or whatever.

> Medicine has become the art of managing extreme complexity. There are over 13,000 different diseases, syndromes and types of injury. Clinicians now have 6,000 drugs and 4,000 medical and surgical procedures. It's a lot to get right.
>
> —ATUL GAWANDE

Simple Solutions

> Ask a typical American hospital what its death and complications rates for surgery were during the last six months and it cannot tell you.
>
> —ATUL GAWANDE

"Count your mistakes." The mere act of counting medical mistakes and errors will help you to identify where to make big improvements without spending a fortune.

> Following the recipe is essential to consistent quality over time.
>
> —CHEF JODY ADAMS

"Make a checklist." You make a list of things to get from the supermarket so that you won't forget something. Why not make a checklist of the things you need to remember for patients? Checklists protect everyone, including the most experienced among us, from making a potentially fatal mistake.

There are two types of checklists:

- Checklists of "stupid but critical stuff" not to be overlooked
- Checklists to ensure that people communicate and coordinate before, during, and after action

Checklists catch mental flaws inherent in all of us—
flaws of memory and attention and thoroughness.
Could a checklist be our soap for surgical care—simple,
cheap, effective and transmissible?

—ATUL GAWANDE

A good checklist shouldn't take any longer than 30 to 90 seconds to complete, or people start taking shortcuts.

Simple Solution: Mistake-Proofing.   www.ahrq.gov/QUAL/mistakeproof/mistakeproof.pdf.

A Better Operating Room in Two Minutes.   The World Health Organization's two-minute Surgical Safety Checklist and implementation manual is available at www.who.int/patientsafety/safesurgery/ss_checklist/en/index.html.

Simple Solutions:
Before incision: Confirm that all surgical team members have introduced themselves by name and role. This helps to create teamwork essential to success.
Preoperatively: "Ask an unscripted question" of the patient, says Atul Gawande. "Where did you grow up?" "How long have you lived in this city?"

Getting the surgical team to introduce themselves and getting patients to talk, even a little, makes it easier for them to speak up when something's not quite right.

Simple Solution: The circulating surgical nurse whose central job is to keep the team antiseptic by retrieving needed instruments and supplies, answering phone calls, doing paperwork, and getting help (Gawande, 2007).

Reduced Infection Rates.   According to the Centers for Disease Control and Prevention (CDC), 2 million Americans acquire an infection while in the hospital.
One-third to one-half of nurses and doctors don't wash their hands consistently.

Simple Solutions:
1. Wash your hands before touching a patient.
2. Ask, "Why can't you wash your hands?" Answers will identify where to add dispensers and other supplies.
3. "Positive deviance"—build on the capabilities people already have rather than asking them to change. Instead of training, say, "Hospital infections are a problem, and we want to know what you know about how to prevent them." Then implement what doctors, nurses, housekeeping, and food service tell you to do.
4. At the Pittsburgh VA Hospital, this approach reduced methicillin-resistant *Staphylococcus aureus* (MRSA) infection rates to zero.

Verifying Results

Purpose: Verify that the problem and its root causes have been reduced.

1. To ensure that the improvements take hold, we continue to monitor the measurements (CTQs). Both the control chart and the Pareto chart will improve if the countermeasures have been successful.
2. Verify that the indicators (CTQs) used in defining the countermeasures have decreased to the target or below (Figure 4.14).

Verify that the major contributor identified in the Pareto chart in defining the countermeasures has been reduced by comparing before and after Pareto charts (Figure 4.15).

To ensure that the improvements take root, we need to develop a flowchart of the improved process and a way to measure its ability to meet patients' needs.

There is always a best way of doing everything.
—RALPH WALDO EMERSON

FIGURE 4.14

Control chart of improvement.

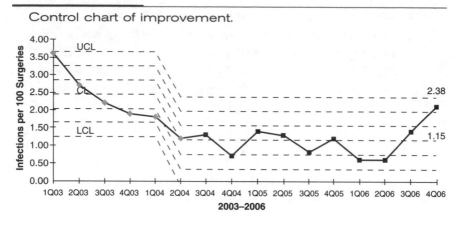

FIGURE 4.15

Comparison Pareto charts showing improvement.

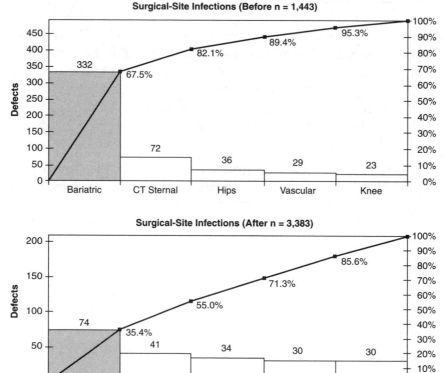

Sustain the Improvement (Control)

Purpose: Prevent the problem and its root causes from coming back.

Like crops in a garden, most improvements will require a careful plan to ensure that they take root and flourish in other gardens. To transplant these new improvements into other gardens will require a plan to monitor and control ongoing performance.

Putting in a "control" system to measure, monitor, and ensure that the improvements stick is the most overlooked step in the improvement process. Sustaining the improvement may require changes to policies, procedures, process flows, and measurements. The QI Macros control plan, flowcharts, and control charts can assist in documenting and monitoring ongoing performance.

What? (Changes)	How? (Action)	Who?	When? (Start	Complete)	Measure? (Results)
People	Training				
Process	Define system and measures				
	Implement				
	Monitor with control charts				
Machines (computers, vehicles, etc.)					
Materials (forms and supplies)					
Environment					
Replicate	Identify areas for replication				
	Initiate replication				

Multiply the Gains

Purpose: To increase the return on investment from each improvement effort.

To maximize your return on investment, you will want to get this improvement into the hands of all the other people who can use it. If you find a way to solve a certain kind of problem in one hospital, replicate the solution in other hospitals.

Where will this process be useful?	What needs to be done to initiate?	How will the process be replicated?	Who owns the replication?	When? Start? Complete?
		Adopt process		
		Adapt process		
		Incorporate existing improvements		

THE 4-50 RULE

In the late 1800s, when Vilfredo Pareto, an Italian mathematician, created what we've come to know as the *80/20 rule*, it changed how we think about life and business, and it laid the groundwork for total quality management (TQM) and Six Sigma.

The Pareto principle is a power law. Instead of numbers spread evenly like butter on bread, Pareto found that 20 percent of the people held 80 percent of the wealth. Because Pareto's rule is a power law, it also applies to itself. This means that as little as 4 percent of people hold 64 percent of the wealth. In America, for example, 4 percent of Americans hold over half the wealth (50 percent).

Pareto also suggested that even if the wealth were redistributed equally to everyone in society, in a very short time it would revert to the 80/20 distribution.

After working with many teams over two decades, I narrowed Pareto's rule to the *4-50 rule*: 4 percent of any business process (one step out of 25) is causing over 50 percent of the waste, rework, defects, and deviation.

SIX SIGMA TAR PITS

Recently, I was called in to facilitate a team that had been in existence for 6 months. All the team had to show for its time was a flowchart of a process that was mainly rework. I'd been calling for weeks, nagging the team for data about how the process performs. I got part of the data the night before the meeting and the rest of the data by lunch. But after a morning of trying to sort through the issues surrounding the process, the team had fallen into "storming" about the whole process. Team members were frustrated, and so was I.

Pitfall 1: *Brainstorming.* Brainstorming is supposed to improve creativity, broaden associations, spark insights, and generate lots of creative ideas. When I first learned TQM, the instructors taught us to brainstorm problems to work on. The difficulty was that we had no idea what a good problem looked like. And it's hard to tell a team to brainstorm a problem to solve and then tell them that their problem is stupid. Lots of teams were started; few succeeded.

This highlights the main problem with brainstorming: If you don't know what you're looking for, you won't get useful ideas.

Brainstorming Research

In the book *Made to Stick* (Random House, 2007), the Heath brothers reference a study of brainstorming. Groups were supposed to create the marketing ideas for a product.

- One group just started creating ideas.
- The second group was given 2 hours of training in brainstorming methods.
- The third group was given 2 hours of training in the six most successful *templates* for ads.

All ads were evaluated by a marketing director and tested on customers.

- Group 1's ads were considered *annoying* by customers.
- Group 2's ads were considered less annoying but no more creative.
- Group 3's ads were considered 50 percent more creative and generated a 55 percent better response from customers.

In other words, brainstorming is useless unless you know what you're looking for or have a template for success.

Solution: When I look for problems to solve with Lean Six Sigma, I'm always looking for something I can solve with the methods and tools.

You can't fix your supplier's or customer's process, which is often the end result of brainstorming; you can see everyone else's faults, but not your own. I can't tell you how many teams try to fix management or their suppliers or their customers. You can't fix someone else's process *because you don't own it*.

You can't fix morale with Lean Six Sigma. You can't fix perceptions. You can, however, fix the underlying problems that lower morale and perceptions.

When it comes to Lean Six Sigma, I'm always looking for

- *Delay.* Sluggish processes always can benefit from the application of Lean. Most of the delay is between process steps when the product is waiting for the next action.
- *Defects.* Error-prone processes devour profits in waste and rework. If detailed numerical counts of defects and their effects (i.e., costs) exist, then it's easy to use Six Sigma's problem-solving process to find and fix the problem. Where there are no facts and figures about the problem, Six Sigma fails.
- *Deviation.* Variation from the ideal target causes higher costs and lower profit margins. Common types of deviation include
  - Too long or too short
  - Too big or too small
  - Too wide or too narrow
  - Too little or too much
  - Too fast or too slow

Get the idea? It's okay to brainstorm problems about one of these three *templates* for improvement, but it's usually worthless to brainstorm without them. Worst of all, most teams hesitate to identify the really pressing problems because they don't want to be on the hook for fixing them.

Tip: Focus on the worst first.

Fix the worst problems, and everything else starts falling into place. Stop majoring in minor things.

Pitfall 2: Starting a team when there are no data (control chart and Pareto chart minimum) indicates that you have a problem that cannot be solved using Six Sigma.

Without data to guide you, you don't know who should be on the team, so you end up with different people trying to solve different problems.

Solution: *Set the team up for success:* (1) Work with data you already have; don't start a team to collect a bunch of new data, and (2) refine your problem before you let a group of people get in a room to analyze root causes.

You can guarantee a team's success by laser-focusing the problem to be solved. One person can do this analysis in a few days using the QI Macros.

Pitfall 3: *Question data.* To throw a team off its tracks, some member who doesn't like the implications of the data will state in a confident voice that the data are clearly wrong. If you let it, this will derail the team into further data analysis. I know from experience that all data are imperfect.

Often the data have been systematically distorted to make the key players look good and to manipulate the reward system, but it is the *systematic* distortion that allows you to use the data anyway.

Solution: Recognize that this member is operating on gut feel, not data. Simply ask, "Okay, do you have better data?" (He or she doesn't.) Then ask, "How do you know the data's invalid?" (I just know.) "How do you know?" (Instinct, gut feel.) "Well, unless you have better data that prove these data invalid, we're going to continue using these data. You're welcome to go get your data, but meanwhile, we're moving forward."

If the person is unwilling to continue, you should excuse him or her from the team because he or she will continue to sabotage the progress.

Pitfall 4: *Whalebone diagrams.* When searching for root causes, if your fishbone diagram turns into a "whalebone" diagram that covers several walls, then your original focus was too broad.

Solution: Go back to your Pareto chart. Take the biggest bar down a level that is more specific. Write a new problem statement. Then go back to root-cause analysis.

Pitfall 5: *Boiling the ocean.* Teams have an unflinching urge to fix big problems or all the problems at once. If you've done a good job of laser-focusing your problem, you'll have a specific type of defect in a specific area on which to focus. If you let the team expand its focus, you'll end up whalebone diagramming and having to go back to a specific problem.

Solution: Get the team to agree to solve just this one issue because its solution probably will improve several other elements of the overall problem. Assure team members that you'll come back to the other pieces of the problem, but first you have to nail this one down.

Pitfall 6: *Measuring activity, not results.* Companies count the number of Six Sigma black belts trained and the number of teams started but fail to measure the results achieved by those teams.

Pitfall 7: *Comparing your results with national averages (i.e., benchmarks).* Just because the national average for turnaround times in the ED is 4 hours, don't start celebrating because your turnaround time is only 3 hours.

Paul O'Neill, former CEO of Alcoa and head of the Pittsburgh Regional Health Initiative (PRHI), says, "The establishment of the idea of national norms is the enemy of continuous improvement." Or, as Jim Collins says, "Good is the enemy of Great."

Here's my point: Use data for illumination, not support. Let the data be your guide. The answers will surprise you and accelerate your journey to Lean Six Sigma.

BECOME A LEAN SIX SIGMA DETECTIVE

In the August 2005 issue of *BusinessWeek*, Michael Hopkins explored the best-seller *Freakonomics* (William Morrow, 2005) and its authors' strategy for using data to explore and explain the world. They wrote: "Morality represents the way that people would like the world to work—whereas economics represents how it actually does work."

The November 2005 issue of *Fast Company* called 2005 the "Year of the Economist." Why? Because books such as *Freakonomics: A Rogue Economist Explores the Hidden Side of Everything*, by Steven Levitt and Stephen Dubner, became a best-seller. Financial columnist Tim Harford says, "The idea of the economist as a detective hero suddenly became easy to sell once *Freakonomics* climbed the best-seller lists."

Suzanne Gluck, the author's agent, says that people are using "freakonomics" as a code word for unconventional wisdom. What's the secret?, *Fast Company* asks. "It's just math," replies coauthor Dubner.

Isn't that the essence of Lean Six Sigma? Using numbers to explore the hidden side of defects, delays, and costs in ways that reveal the hidden gold mine of profits wasted every day in businesses large and small.

What's the "secret sauce" that makes Steven Levitt so successful? Coauthor Dubner says, "He seemed to look at things not so much as an academic but as a very *smart and curious explorer*—a documentary filmmaker, perhaps, or a forensic investigator or a bookie whose markets ranged from sports to crime to pop culture." *He is an intuitionist.* He sifts through a pile of data to find a story that no one else has found. The *New York Times Magazine* said he's "a kind of intellectual detective trying to figure things out."

Isn't that what Lean Six Sigma is at its core? Sifting through piles of data like an intellectual detective trying to explain the hidden side of defects, delay, and cost?

Solution: The Data Strategy

In Hopkins' article, he identifies the key strategies used by Steven Levitt and Stephen Dubner. They are

1. *Use your data.* Experts use their informational advantages to serve their own agendas. Hence the numbers can be bent to prove whatever they want to prove. It's amazing how many company managers want to use data to prove their pet theory or justify their actions. Only after a long struggle do they begin to learn how to use data as a guide to clear thinking and action. I've always liked the quote: "He uses statistics like a drunk uses a lightpost, for sup-

port not illumination." In Lean Six Sigma, this holds true far too often.

2. *Knowing what to measure and how to measure it makes a complicated world much less so.* If you learn how to look at data in the right way, you can explain riddles that otherwise might have seemed impossible.

   In the book *Moneyball* (W. W. Norton, 2003), Michael Lewis explains how the Oakland A's used statistics to consistently pick excellent but undervalued players for the team. In *Think Twice* (Harvard Business Press, 2009), Michael J. Mauboussin explains how Orley Ashenfelter, a wine enthusiast, used regression analysis to explain and predict the quality (and price) of Bordeaux wines using rainfalls and temperatures.

   Companies generate lots of data about orders, sales, purchases, payments, and so on. The bigger the company, the more data it has and the less likely it is to use those data. Figure out what data are useful and *use* them. Figure out what data aren't useful, and stop collecting them.

3. *Ask quirky questions.* If you're focused on why things go wrong, ask, "What are we doing right? Who is already doing this right?" If you focus on why things are going right, focus on what's wrong and start with the "worst first." In *Freakonomics*, Levitt and Dubner stopped asking why crime rates have fallen since 1990. They started asking what kind of individuals are most likely to commit crimes and then asked, "Why are they disappearing from the population?" Their answer to this question is startling but instructive of their method: "Let the data lead you."

4. *Don't mistake correlation for causality.* America spends 2.5 times more on healthcare than any other country, yet Americans aren't healthier than other people. Affluent women have a higher incidence of breast cancer than poor women. Does wealth cause breast cancer? Does healthcare cause illness?

5. *Dramatic effects often have distant, even subtle causes.* Six Sigma looks for direct cause-effects, but systemic effects can amplify subtle causes into dramatic ones.

6. *Question conventional wisdom.* The conventional wisdom is often wrong. If conventional wisdom were correct, then

most problems would have been solved already. You can't get new insights from old ways of thinking.

7. *Respect the complexity of incentives.* Incentives are the cornerstone of modern life. People are rewarded for following systems that cause defects, delay, and cost. Humans always will find ways to beat the system. Rely on it.

The moral of the story: "Make data your friend," says Hopkins. I'd say, "Let it be your guide."

> **Tip: Follow your data, not your hunches.**

In the book *Freakonomics*, Levitt confesses, "You don't need to know a lot of math. I'm horrible at math." Lots of people are horrible at math. Get over it. This is why I created the QI Macros SPC Software for Excel. The macros do all the scary math; you just need to know how to interpret the resulting graphs.

If you go to www.qimacros.com/hospitalbook.html, you can download the QI Macros Lean Six Sigma Software 90-day trial. Become a Six Sigma detective or treasure hunter superhero. Learn how and what to measure to simplify understanding your business. Let your measurements lead you to find and plug the leaks in your cash flow. Distrust conventional wisdom. Look for subtle causes that amplify themselves into disturbing effects. Share what you learn. Most of all: *Get on with it!* There's no end to the mysteries to be revealed and problems to be solved.

MISTAKES, DEFECTS, AND ERRORS

At the Institute for Healthcare Improvement (IHI) conference in Orlando, Florida, one of the presentations covered the application of the Toyota Production System (TPS) to a hospital. The presenter opened by saying that *healthcare, in general, was a poor-quality product that cost too much for the value delivered*. I was immediately struck by the guts it took to make that statement. The presenter went on to repeat that thought many times throughout the presentation. I doubt that many people caught it, however.

One of the biggest challenges to Lean Six Sigma is not use of the methods or tools but *creating a mind-set that loves to find and fix defects and deviation*. Not everyone thinks of these issues under the banner that I call *defects and deviation*, so I got into the Synonym

Finder to look for other words that mean the same thing. It's amazing how many words exist in the English language to describe mistakes and errors. Here are just a few:

blemish	fallacy	misprint
blooper	false step	misstep
blot stain	fault	mistake
blotch	faulty	muff
blunder	flaw	off the beam
bobble	flub	omission
boner	foul-up	oversight
boo-boo	fumble	rough spots
botch	goof	scare deformity
breach	human error	scratch
bugs	illogical	screw-up
bungle	imperfection	shortage
clinker	imprecise	shortcoming
clunker	inaccuracy	slip up
cockeyed	inadequacy	snafu
crack	incomplete	snags
defect	incorrect	spot
deficiency	inexact	tear
drawback	kinks	trip
error	leak	unsound
failing	louse up	weak point
failure	miscue	weakness

If you continue to look at words that describe how people make these mistakes, you'll find another group of words dedicated to describing *the activities that lead to poor-quality products and services.*

misapply	misdoing	mismanage
misapprehend	misestimation	mismatch
miscalculation	misguided	misplace
misconceive	mishap	misreading
misconception	misidentification	misreckon
misconstruction	misinterpretation	misspend
misconstrue	misjudgment	misstep
miscount	mislay	mistaken
misdirected	mislead	misunderstanding
		misuse

Until you're willing to *stop congratulating yourself for what's working and start looking at the misses, mistakes, errors, omissions, defects, and delay* that are irritating patients, demotivating employees, and devouring your profit margins, all the Lean Six Sigma methods and tools will not help you. Once you view every mistake as an opportunity to mistake-proof and improve the delivery of healthcare, you'll get hooked on Lean Six Sigma. Until then, the methods and tools will just be another burden in an already crisis-managed world.

MEASUREMENT SIMPLICITY

Jack Welch said, "Simplicity applies to measurements also. *Too often we measure everything and understand nothing.*" All too often I hear from QI Macros users that they are so overwhelmed drawing charts and graphs for management and other people that they don't have time to analyze and improve anything. This is what *theory of constraints* considers to be a "dummy constraint"; if everyone had software like the QI Macros, they could draw their own charts and free the process-improvement people to focus on improvements, not charts.

One hospital using the QI Macros was tracking 300 different measures. 300? What's wrong with this picture? I'll tell you what— They can't possibly be using all these measures. From a Lean standpoint, this is classic overproduction. Ten or 12 measures provide most of the information required to run that hospital. Measurements should help you, not hinder you.

The purpose of measurement is to guide, forewarn, and inform.

1. Guidance provides course corrections "in flight" while you're running the business.
2. Measurement also can forewarn you of potential problems (e.g., trends or instabilities on control charts).
3. And measurement can help keep customers, suppliers, and leaders informed of your progress.

Are you collecting measurements that aren't really useful for any one of these three purposes? Do you really need them? Is some other measurement used in their place? (4-50 rule: 4 percent of the measurements cover 50 percent of your needs.)

First, start systematically suspending measurements that are questionable. Then, if anyone comes out of the woodwork to complain about missing the information, ask, "How are they using the information? Would some other measurement serve them better?" Second, if a suspended measurement isn't resurrected in 2 or 3 months, kill it. Third, start looking for the "vital few" measurements of "failure" that everyone relies on to make improvements and informed decisions. In any business these are invariably defects, delay, and cost. You'll also need measurements of success: profit, return on investment, and so forth.

Here are four basic steps to create your own process measures:

1. Define what results are important to you and the hospital.
2. Map the cross-functional process used to deliver these results.
3. Identify the critical tasks and capabilities required to complete the process successfully.
4. Design measures that track those tasks and capabilities.

What are the most common measurement mistakes?

1. *Piles of numbers.* Use the balanced scorecard (one of the QI Macros templates) to identify the vital few.
2. *Inaccurate, late, or unreliable data.* If it isn't collected systematically and automatically in real time, it's often suspect.
3. *Trying to meet a target instead of trying to understand the process.*
4. *One size fits all.* Trying to use too broad or too specific a measurement.
5. *Gauge blindness.* Trusting the measurement even when there is evidence to the contrary (e.g., a sticky gas gauge can leave you stranded).
6. *Micrometer versus yardstick.* Precisely measuring "unimportant" things without imprecisely measuring the "important" ones.
7. *Punishing the people instead of fixing the process.* Use your data to learn something and make processes better.
8. *Simplify and streamline your measurement system* to keep the important stuff and to abandon the unimportant stuff. You'll be surprised how much unimportant stuff is sucking up time and resources that could be dedicated to improving your business!

ACCIDENTS DON'T JUST HAPPEN

This is the sort of headline that you don't want to read about your hospital:

> **Accident Kills Boy Undergoing MRI**—A 6-year-old boy was killed when the MRI's powerful magnet pulled a metal oxygen tank through the air, fracturing the boy's skull. Westchester Medical Center officials said the tank had been brought into the room *accidentally*. Officials would not say who brought the oxygen tank into the MRI room.

Forget *who* brought it into the room. Doesn't it seem more desirable that it should be *impossible* to bring metal into an MRI room? I've been through scanners at airports that rant about anything bigger than a quarter. Could such a device be installed in the doorway to the MRI? Sure. Could an alarm on the metal detector prevent operation of the MRI? Sure. If it saves the life of one 6-year-old boy, wouldn't it be worth it?

Now ask: Why? Why? Why? Why? Why? Why were oxygen tanks loaded or unloaded anywhere near the MRI? Is the MRI close to the loading dock? Was the boy brought in from surgery with a tank? Why wasn't his tank removed before the MRI?

ANALYSIS IS EASY . . . IF YOU KNOW WHAT TO LOOK FOR

The process is simple:

1. *Look for data on defects, mistakes, or errors over time.* Draw a control chart of performance over time.
2. *Draw a Pareto chart of the types of defects found.*
3. *Use the biggest bar of the Pareto chart to create a fishbone diagram for root-cause analysis.* The problem statement should reflect the problem identified in the Pareto chart. You now have enough insight into the problem to choose the right root-cause analysis team.
4. *Analyze the root causes and verify that you have found the true root causes using data.*
5. *Show performance before and after implementing the improvement using a control chart.*

6. *Continue to monitor and improve the process.*

I don't know why, but most people try to make it a lot harder than this. You don't have to. You can let your data lead you to dramatic improvements.

CHAPTER 5

Simple Steps to a Cheaper (More Profitable) Hospital

It should come as no surprise that a *faster, better hospital* will be *cheaper to operate and more profitable*. When you're not dealing with all the delays in the emergency department (ED)–admission–discharge process, fewer patients will be boarded in the ED, reducing diversion and patients who leave without being seen (LWOBS). More patients can be seen more quickly, increasing revenue. When you're not dealing with the extra costs of preventable falls, infections, and medication errors, it will make the hospital more cost-effective and profitable.

FASTER + BETTER = CHEAPER AND MORE PROFITABLE!

One year, the National Association of Healthcare Quality conference in Orlando, Florida, sandwiched itself in between Hurricane Ivan and Tropical Storm Jeanne. This was after hurricanes Charles and Frances had pounded the state. Mounds of debris still littered the suburbs.

Although most hurricane seasons deliver three or four hurricanes, most people now have come to think of Florida as the hurricane state, not the sunshine state. What was different about that particular season from the last? It was not the number of hurricanes, because that's about the same as every season. The difference was that they all happened in such close proximity that we were able to detect the pattern: tropical storm grows into a hurricane that hits Florida, the panhandle, or the Gulf Coast.

This is the challenge faced by all quality professionals. It's easy for people to detect mistakes and errors when they happen often enough. But when the frequency of those mistakes falls below a certain level or they have a minor impact, you can no longer detect them with your five senses. You need some better tools. Fortunately, the process and tools are simple:

1. Count your misses, mistakes, and errors.
2. Categorize your misses, mistakes, and errors.
3. Fix the biggest category first.

Count Your Misses, Mistakes, and Errors

While much of healthcare quality is focused on clinical care and the Joint Commission and National Database of Nursing Quality Indicators measures, there's a lot that can be done with the financial or "transactions" side of the business to eliminate costs and increase profits.

One healthcare client had 37,000 rejected insurance claims worth millions in appealed and denied claims. In case it isn't obvious, there was a lot of rework involved in fixing those rejects and appeals. A hospital can have great clinical success but terrible financial issues. Both the clinical and operational sides of the hospital have to work flawlessly to reduce costs and maximize patient satisfaction and outcomes.

Categorize Your Misses, Mistakes, and Errors

My team began by grouping the errors into the categories of rejects, appeals, and denied, but to make these categories actionable, we had to dive a little deeper.

Denied Claims.   Since "denied" claims involved *real* dollars, we did a number of Pareto charts to look for more important categories. The biggest category of denied claims was for those denied for lack of timely filing (within 45 days). Then we categorized the denied claims by insurer. When we did, there was a big surprise: One small insurer accounted for 64 percent of denied claims.

This brings us to back to what I call the *4-50 rule*: 4 percent of the categories cause 50 percent of rework, waste, and lost profit. If you want to plug the leaks in healthcare quality, you have to use your data to find and fix these small but costly categories.

Armed with these data, we instituted process changes over the weekend that started saving $380,000 a month.

Rejected Claims. Rejected claims are bounced at the doorway of the insurer owing to incorrect or incomplete information. When we looked at the rejected claims data, one category, "duplicate claims," stood out as 24 percent of the problem, or about $35 million. We used what I call the "dirty 30 process" for improvement: We looked at 30 examples of duplicate claims and discovered that these were simply miscoded. The claim had been paid; there wasn't a duplicate claim. By changing the procedures and using the correct code, we eliminated a problem that inflated the overall error numbers.

Other categories included such things as the alpha prefix on Blue Cross Blue Shield insurance ID cards and uninsured dependents such as newborns and students.

Fix the Biggest Category First

Therefore, category by category, we used the "dirty 30 process" to identify and fix each of the most common problems first. These fixes often reduced other related categories as well. While most teams want to boil the ocean or solve world hunger, when you restrict yourself to fixing the biggest category first, you'll find it easier to make a difference, and a surprising amount of benefit will come along with it.

Appeals. In the third big category, appeals, the real problem was time. The average length of time for an appeal to be resolved was 298 days. By focusing on the cycle time, my team was able to get the time under 90 days, which, of course, improved cash flow by over $800,000 and enhanced the bottom line.

Don't Wait for Hurricane Season

Every healthcare business is plagued by occasional hurricanes that are fanned by a series of problems that happen frequently enough to penetrate the dense fog of consciousness, but why wait? You already have the data you need to start finding and fixing the major categories of misses, mistakes, and errors in all aspects of your facility.

Don't just focus on the clinical side; turn your attention to the big problems on the transactions side of the house: purchasing, billing, and claims. A recent study estimated that *eight out of ten hospital bills have mistakes*. There's a whole class of consultants who, for a fee, will help patients to navigate this maze to get their bills paid. *Note:* When this happens, your patients have become part of your "fix-it factory."

It's not enough to provide high-quality clinical care; you also must provide a high-quality experience in all aspects of service delivery from admissions to discharge and beyond. Patients are no longer patient. The global marketplace has taught them the value of better, faster, and cheaper. Consider thinking of them as "impatients."

To reduce rejected, appealed, and denied claims, use Six Sigma tools to focus the improvement effort. The process is simple:

1. Analyze each type of claim or defect using control charts and Pareto charts:
   a. Rejected claims
   b. Appealed claims
   c. Denied claims
2. Analyze the root causes using the "dirty 30 process."
3. Implement countermeasures.
4. Track results.

REDUCING DENIED CLAIMS IN FIVE DAYS

Denied claims mean no money for services rendered because the billing process failed in some way. Nonpayment drives up the cost of healthcare and pushes many hospitals toward bankruptcy. In this case study, monthly denials were over $1 million (Figure 5.1).

Using Pareto Charts of Denials

Using Excel PivotTables and the QI Macros, it was easy to narrow the focus to a few key areas for improvement: timely filing (Figure 5.2) and one insurer (Figure 5.3).

FIGURE 5.1

Control chart of denied claims.

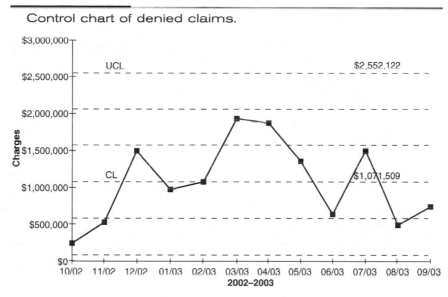

FIGURE 5.2

Denied claims–timely filing Pareto chart.

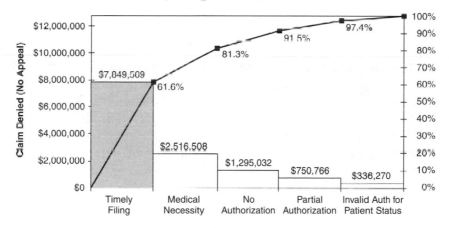

Analyze Root Causes and Initiate Countermeasures

In a half-day root-cause analysis session, my team identified ways to change the process to work around the denials and change the

FIGURE 5.3

Timely filing by payer Pareto chart.

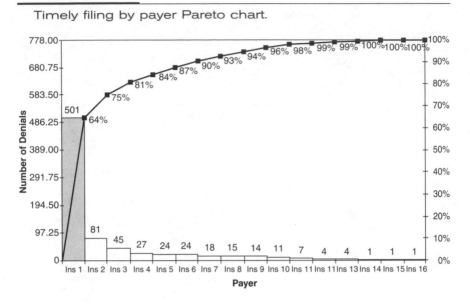

contract process to (1) reduce delays that contribute to timely filing denials and (2) work with the insurer to resolve excessive denials.

Verify Results

After we implemented the process changes the following Monday, denied claims fell by $380,000 per month ($15 million per year). An XmR chart (Figure 5.4) shows denials before and after improvement.

If you go to www.qimacros.com/hospitalbook.html, you can download the QI Macros Lean Six Sigma Software 90-day trial. Use the PivotTable Wizard to create pivot tables. Use the Control Chart Wizard to draw control charts. Use the Pareto chart macro to draw Pareto charts.

REDUCING REJECTED CLAIMS IN FIVE DAYS

In software we have a saying: "Finding a bug in a computer program is like finding a cockroach in your hotel room. You don't say: 'Oh, there's a bug.' You say: 'The place is infested.'" The same is true of rejected claims. Start with a line or control chart of rejects (Figure 5.5).

FIGURE 5.4

Denied charges improvement control chart.

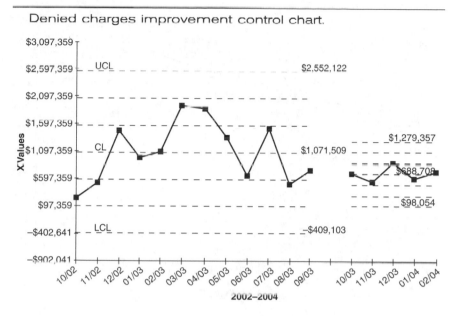

FIGURE 5.5

Rejected, appealed, and denied claims line graph.

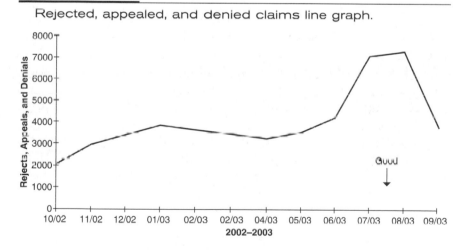

Use a Series of Pareto Charts
to Narrow Your Focus

Rejected claims are the most frequent type of error (Figure 5.6); appeals tie up accounts receivable, and denials result in lost revenue. How can we use Lean Six Sigma? Start with rejected claims.

FIGURE 5.6

Rejected claims Pareto chart.

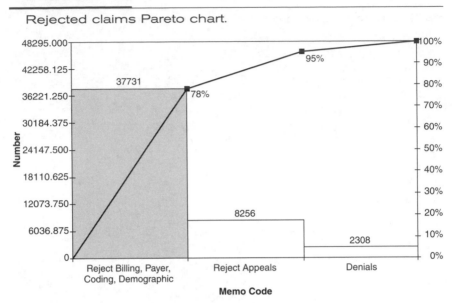

Categorize Rejected Claims

Duplicate claims account for 27 percent of rejected claims (Figure 5.7). The next four bars of the Pareto chart, combined with duplicate claims, account for 80 percent of all rejected claims. Each of these five bars of the Pareto chart is an improvement story requiring root-cause analysis.

Let's take duplicate claims down to the next level of the Pareto chart (Figure 5.8).

In this example, secondary payments for Medicare patients account for 83 percent of the duplicate claims. The team investigated 72 of these secondary payments and found that they had been paid but were coded incorrectly in accounting. A simple process change reduced duplicate claims by $24 million.

Team Continued with the Other Four "Big Bars"

- "No coverage" turned out to be caused by charges after policy termination (Figure 5.9).
- "Invalid insurer information" led to "Social Security Number incorrect" and "wrong primary insurer" (Figure 5.10).

FIGURE 5.7

Rejected duplicate claims Pareto chart.

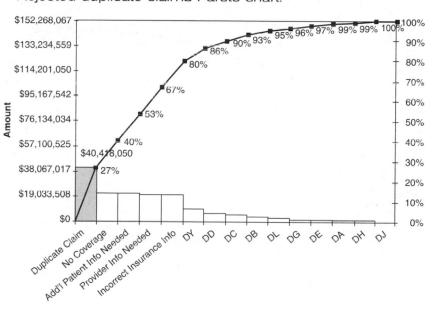

FIGURE 5.8

Duplicate claim verification Pareto chart.

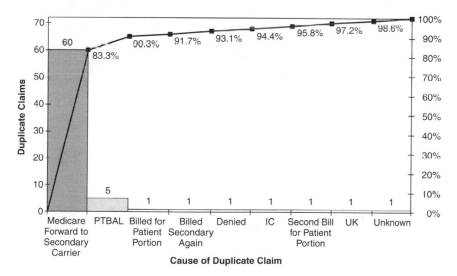

FIGURE 5.9

No coverage Pareto chart.

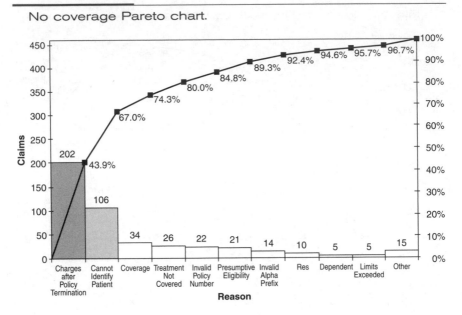

FIGURE 5.10

Invalid insurance information Pareto chart.

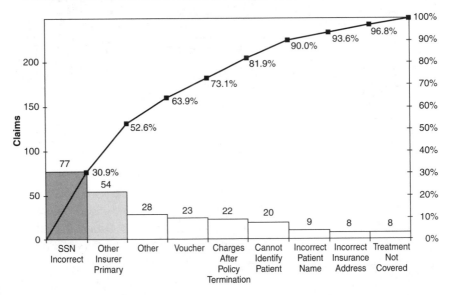

FIGURE 5.11

Rejects for patient information errors Pareto chart.

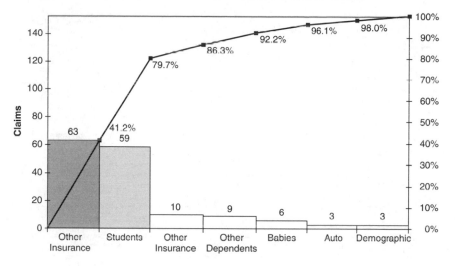

- "Patient information" rejects (Figure 5.11) led to analysis of "Other insurance" (41 percent) and "Students missing from parent's insurance" (39 percent).

Results

Half-day root-cause analysis sessions for each of the "big bars" on these Pareto charts and subsequent improvements resulted in dramatic improvement in "first-pass yield" of insurance claims.

- 72 percent reduction in emergency department (ED) billing rejects
- 60 percent reduction in impacted charges

REDUCING APPEALED CLAIMS IN FIVE DAYS

Delayed payments caused by appealed claims can put a hospital in a financial crunch. In 2003, appealed claims spiked owing to Medicare Part B changes (Figure 5.12). The recent healthcare reform legislation and subsequent changes most likely will cause further spikes.

FIGURE 5.12

Reject appeals control chart.

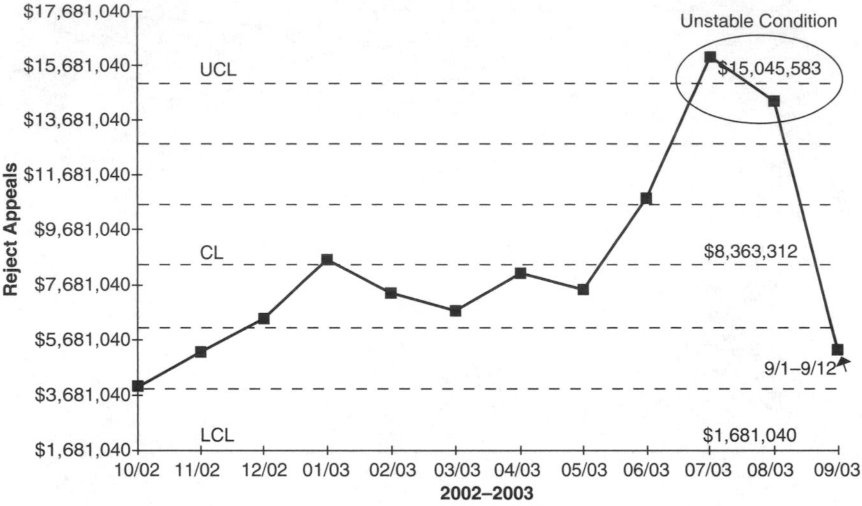

Use Pareto Charts to Analyze Appealed Claims

There are a number of ways to analyze appeals data, for example, by patient and by appeal type (Figures 5.13 through 5.15).

From these Pareto charts, authorization and precertification of admissions from the ED are the most common and costly appeals. Root-cause analysis required team members from the ED and admissions to identify and reduce authorization/precertification appeals.

Reducing Cycle Time for Appealed Claims

Turnaround time for appealed claims can be shown in a histogram (Figure 5.16). Using the simple tools of Lean (Post-it Notes), it was possible to redesign the appeals process to

- Reduce touches per account from 21 down to 11
- Reduce minutes per touch (16 minutes)
- 2.97 hours saved per account (estimate of savings using loaded rate of $50/hour gives $150 per account)
- Accelerate payment by 50 days

FIGURE 5.13

Rejected appeals Pareto chart.

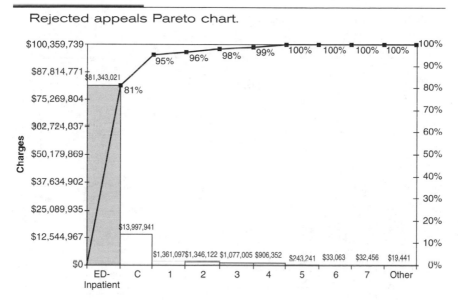

FIGURE 5.14

Appealed claims Pareto chart.

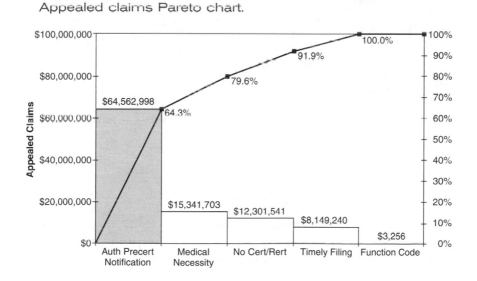

FIGURE 5.15

Authorization/precertification appeals by patient-type
Pareto chart.

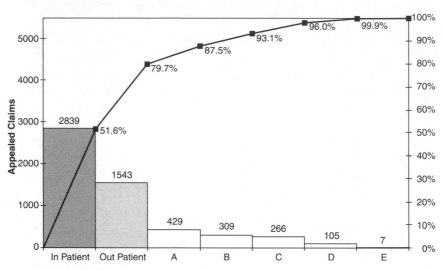

FIGURE 5.16

Appeals delay histogram.

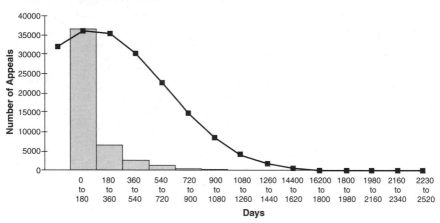

Other Examples

In 2002, rural hospital Thibodaux Regional Medical Center used
Lean Six Sigma to reduce "discharged not final billed" from $3.3
million to $600,000. They also reduced net accounts receivable from
73 to 62 days, resulting in an increased cash flow of $2 million per

year. A second wave of projects saved an addition $489,000 per year in inventory costs.

In 2006, North Shore Long Island Jewish Health System used Six Sigma to reduce oncology billing errors (missing charges) from 50 percent to only 2.5 percent. This resulted in increased revenue of $4 million per year. The system also reduced turnaround time for charge entry from 3.7 to 2.4 days and DOS-to-billing from 13.6 to 6.8 days. The biggest factor in timely filing of bills? Missing information. The biggest culprit? The pharmacy (Figure 5.17).

United Healthcare worked with one hospital group to reduce unreconcilable claims from 23 to 4.8 percent in 6 months.

HOW TO GET A CHEAPER (MORE PROFITABLE) HOSPITAL IN FIVE DAYS

From working with teams in various industries, I've developed a simple method for achieving breakthrough improvements on transactions processes such as billing. I call it the "dirty 30 process." I used it in the case study presented in this chapter.

FIGURE 5.17

Missing billing coding Pareto chart.

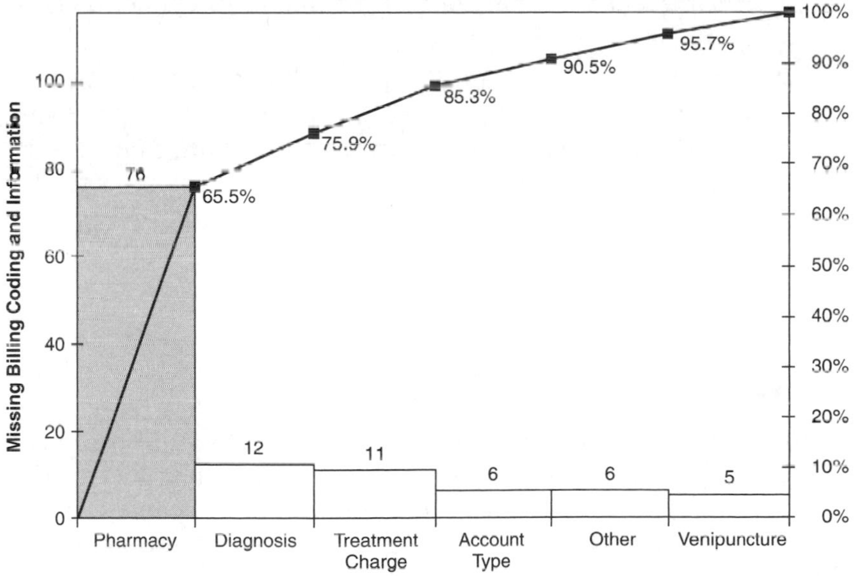

The "Dirty 30 Process" for Better Billing

The secret is to

1. Quantify the cost of correcting these rejected, appealed, and denied transactions.
2. Understand the Pareto pattern of rejected, appealed, and denied transactions.
3. Analyze 30 to 50 of each main type of rejected, appealed, or denied transactions to determine the root cause.
4. Revise the process and system to prevent the rejected, appealed, or denied claim.

Process: Typical root-cause analysis simply does not work because of the level of detail required to understand each error. Detailed analysis of 30 errors in each of the top error "buckets" (i.e., the "dirty 30") led to a breakthrough in understanding of how errors occurred and how to prevent them. Simple checksheets allowed the root cause to pop out from analysis of this small sample. As expected, the errors clustered in a few main categories.

The "dirty 30 process" has four steps:

1. *Focus.* Determine which rejected, appealed, or denied error buckets to analyze first for maximum benefit. (This analysis takes 2 to 3 days.)
2. *Improve.* Use the "dirty 30" approach to analyze root causes (4 hours per error type—using a facilitator with the team), and determine the process and system changes necessary to prevent the problem.
3. *Sustain.* Track the rejected, appealed, and denied claims after implementation of the changes.
4. *Honor.* Recognize and reward team members.

INSIGHTS

Using the basic tools of Six Sigma, anyone can learn to use what I call the "dirty 30 process" in a day or less to find the root causes of transactions errors. Once a team has found the root causes of these errors, it's just a matter of changing the processes and systems to eliminate the errors.

Hundreds of people spend their lives fixing the fallout from these rejected, appealed, and denied claims. And they all think that they're doing meaningful work, not just fixing things that shouldn't be in error.

CONCLUSION

Until you get to where you can prevent errors, every system could benefit from a simple yet rigorous approach to analyzing and eliminating errors. The "dirty 30 process" is ideal because the data required to implement it are collected by most systems automatically. Then all it takes is 4 to 8 hours of analysis to identify the root cause of each error.

CHAPTER 6

Six Sigma for Hospitals

Every hospital has two "factories":

1. *A "good" factory that delivers patient care.* In a hospital, this would be the emergency department (ED), operating room (OR), nursing units, lab, radiology, pharmacy, admissions, billing, and so on.
2. *A hidden but expensive "fix-it" factory.* This is the factory that cleans up all the medical mistakes, operational mistakes, and delays that occur in the main factory.

Hospital EDs experience repeat visits from the same patient in the same day. Doctors misdiagnose illnesses 15 percent of the time, according to one *Business Week* article. If your hospital is typical, the fix-it factory is costing you $25 to $40 of every $100 you spend. If it costs $100 million to run your hospital, you spend $25 million to $40 million on the fix-it factory.

DOUBLE YOUR PROFITS

Most hospitals run on razor-thin margins of less than 5 percent ($5 million in a $100 million hospital). Reducing delays, defects, deviation, and costs by 20 percent ($20 million) would more than double your profits. Just think what saving a fraction of that waste could do for your productivity and profitability!

The urgencies of any business can consume all your time. Fortunately, given the right gauges on your dashboard, it's easy to diagnose where to focus your improvement efforts, even while you are still working in your business.

The End of Common Sense

When I worked in a phone company, managers used to say that process improvement is "just common sense," but what I've learned is that common sense will only get you to a 1 to 3 percent error rate. Hospitals get to a 1 percent error rate on such things as infection rates and medication errors, but that's where they reach the edges of human perception, *the end of common sense*.

When you reach the end of what you can do with one problem-solving technology (e.g., common sense), you need to look to the next level: systematic problem solving and the tools of Lean Six Sigma. The primitive methods and tools that took you to sustainable profitability will take you no further. To turn your cash cow into a golden goose, you will need the *common science* in Lean Six Sigma to make breakthrough improvements. Here's what you can accomplish with Lean Six Sigma:

1. *Double your speed without working any harder.* Most hospitals have extensive delays *built into* their processes. Eliminate the delays and you can run circles around your competition and delight patients.

2. *Double your quality* by reducing defects and deviation by 50 percent or more. Lean alone has been shown to reduce defects by 50 percent. The IHI estimates that one out of every two patients suffers some preventable harm. Lean could help to lower that rate to one out of three. Add Six Sigma, and you've got a recipe for world-class performance.

3. *Cut costs and boost profits* because every dollar you used to spend fixing problems now can be refocused on growing the business or caring for patients or passed right through to the bottom line. Instead of wasting 25 to 40 percent of every dollar you spend fixing things that shouldn't be broken, most of that money can fall through to the bottom line, boosting margins through the roof.

MANUFACTURING AND SERVICE

At an abstract level, there's no real difference between a service process and a manufacturing one. They both encounter delays, defects, deviation, and costs. One may produce discharged patients instead of disk brakes, purchase orders instead of computers, bills instead of brake liners, but they all take time, cost money, create defects, cause rework, and create waste.

In a hospital, we might focus on medication errors. We might focus on variation in admission, diagnosis, treatment, or discharge delays. We might focus on the costs of medical errors that result in longer hospital stays.

In a hospital, the clinical side is only one element. Defects and delays in issuing bills and insurance claims can cost millions of dollars. This is true in any company, from a family-owned restaurant to a Fortune 500 company. Incorrect bills, missing charges, incorrect purchase orders, overpayments, underpayments, and so on can cost a fortune. Fielding the phone calls and fixing the financial transactions can cost more than some invoices are worth.

Ken Miller, author of *We Don't Make Widgets* (Governing Books, 2006), has identified the three common excuses that employees use for why they can't apply Lean Six Sigma to their service business:

1. *We don't make widgets.* We do "squishy, intangible stuff" like caring for patients.
2. *We don't have customers.* Healthcare: We have patients. (As Ken says, "You don't have patients; you have hostages.")
3. *We're not here to make a profit.* Healthcare: We're here to care for the sick.

Listen to any hospital employee try to tell you why Lean Six Sigma won't work, and you'll usually hear one of these three excuses. Sadly, Ken says that these myths can stop an organization from embracing radical improvement. Once you see healthcare as a system of processes done by people to produce a "widget" (e.g., surgery, insurance claim, bedside meal) to achieve a desired outcome (i.e., healing), you'll be well on the way to understanding how to use Lean Six Sigma to create perfect outcomes for all patients.

TRICKS OF THE TRADE

After a meal at a local Chinese restaurant, my fortune cookie said, "If you keep too busy learning the tricks of the trade, you may never learn the trade." When I think about how this applies to Lean Six Sigma, it seems obvious that far too much Lean Six Sigma training is dedicated to the tricks of the trade and not enough to the actual trade.

The Long Tail of Six Sigma Tools

To fill the long weeks of Six Sigma training, most trainers cover every tool in the toolbox as if they are all equally important. One trainer admitted that his black belt training for healthcare included three days of design of experiments (which healthcare rarely needs).

A 2003 study by *Quality Digest* magazine confirmed what I've known for years: *A handful of tools and methods is delivering most of the benefit from Lean Six Sigma.* Focused application of these tools will carry you from average to excellent in as few as 24 months *while delivering staggering improvements in productivity, profits, and patient care.*

In any profession, there is a handful of tools that are used all the time and a slew that are used once in a very long while. This is true of Six Sigma. There is a "long tail" of tools (Figure 6.1).

FIGURE 6.1

The long tail of Lean Six Sigma tools.

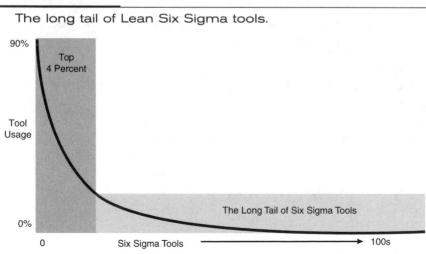

Master the Top Four Percent

One of the principles of adult learning is that participants must use what they've learned in 72 hours or they lose 90 percent of what they've learned. Most Six Sigma training is done in a week-long format. This means that by Thursday, participants have forgotten Monday; by Friday, they've forgotten Tuesday; and by Monday of the following week, they've forgotten most of the previous week.

I know what a lot of trainers are thinking: "But we have case studies they do in class." I have found that unless people apply Six Sigma methods and tools to their own work environment, it just doesn't stick. Classroom case studies are nice, but they won't add money to the bottom line. This should be unacceptable.

A handful of tools (the top 4 percent), such as control charts, Pareto charts, and fishbone diagrams, will solve 90 percent of common problems with defects. If there's a lot of variation (I call it *deviation* because *variation* sounds too benign), throw in a histogram or two.

Master these tools first. Then add the long tail of tools as needed. (Admit it; this is how your home toolkit grew—from a hammer, a screwdriver, and a pair of pliers into a toolbox of gadgets.)

Use company data, not case-study data, to tailor the learning to the participants. Using the company's data, these tools can be learned and applied in a day, not a week.

If you go to www.qimacros.com/hospitalbook.html, you can download the QI Macros Lean Six Sigma Software 90-day trial. Use the PivotTable Wizard to create pivot tables. Use the Control Chart Wizard to draw control charts. Use the Pareto chart macro to draw Pareto charts. Use the Ishikawa diagram to create fishbone diagrams of root causes.

Stone Age and Space Age Tools

Gut feel, trial and error, and common sense are the primitive tools of the Kalahari or Outback, not the tools of operational excellence. Similarly, futuristic tools invented by statisticians hoping to leave their mark on the world are inhibiting the spread of Lean Six Sigma.

There's a handful of methods and tools that will solve most common problems in healthcare. Isn't it time to make it easy for the primitive business tribes to embrace these tools?

What do I think are the essential tools? The ones I use most of the time include

- XmR control charts to show performance over time
- Pareto charts to identify improvement opportunities
- Histograms to analyze deviations from target
- Ishikawa (fishbone) diagrams to show cause-effects
- Value-stream maps to identify delays between process steps

The QI Macros provide easy, affordable access to these tools (download your 90-day trial from www.qimacros.com/demystified.html). Everything else is overkill for the primitive corporate tribes. Get them comfortable and feeling safe in the use of these tools, and then you can begin to add in the others.

Learn the Trade, Not the Tricks!

In school, we learned reading, writing, and arithmetic, but we didn't learn calculus right out of kindergarten. We shouldn't expect employees to skip grade school and start college, but this is what we've done with Six Sigma by covering the long tail of tools—tools employees will rarely use.

Let's start teaching people how to solve common problems in their businesses. Let's make them successful with the top 4 percent of tools and then add the long tail as needed. I run into people I trained 15 years ago who are still using these tools in whatever job they've taken. When they pull their improvement stories out of their desks to show me, I feel good knowing that the one-day training stuck in their skull, took root, and flourished.

Let's teach them the trade—and then the tricks. It's hard to make money in a training business this way, but it's a powerful way to deliver bottom-line savings and boost profits.

THE FASTER, BETTER, CHEAPER TOOLKIT

Lean Six Sigma is the best toolkit for helping you to *think outside the business*. The tools are designed to help employees see the business more clearly than ever before.

Lean Six Sigma is a results-oriented, project-focused approach to quality, productivity, and profitability. These reductions translate into cost savings, profit growth, and competitive advantage. And the process is simple:

1. *Focus* on key problem areas by counting and categorizing your delays, defects, misses, mistakes, errors, and deviation.
2. *Improve* by eliminating delays, defects, and deviation.
3. *Sustain* the improvement by monitoring key measures and responding if they become unstable and unpredictable.
4. *Honor* your progress.

If we applied Lean Six Sigma to hospitals, there would only be 3.4 deaths per million hospital admissions instead of 1 per 300, as reported by the National Academy Press (1999).

If you go to www.qimacros.com/webinars/webinar-dates .html, you can sign up for a free Lean Six Sigma for Healthcare Webinar to learn how to apply Lean Six Sigma in hospitals.

EVERY BUSINESS HAS TWO SOURCES OF CASH FLOW

Cash is the lifeblood of any business. To boost profits, you will want to earn more or lose less. Every business has two sources of cash flow:

1. *External payers* give you money for your patient-care services.
2. *Internal processes* leak cash like a rusty bucket. Why are internal processes a source of cash? Because when you plug the leaks in your cash flow, you get to keep all that money! And it's a lot of money—a third or more of your expenses.

I'd like you to consider that healthcare spends most of its time and money focused on the latest technology for patient care and virtually no time on plugging the money leaks caused by internal processes.

You have complete control of the processes and technology inside the walls of your facility. Every process leaks cash. Even if you only make one mistake in every 100 transactions—orders, bills, purchase orders, payments, products, or services—that 1 percent error rate can add up to 6 to 12 percent or up to 18 percent across the hospital or facility.

The Juran Institute has found that the cumulative cost of delays, mistakes, rework, and scrap will add up to 25 to 40 percent

of your total expenses (Figure 6.2). Don't believe it's that much?
Spend a day tracking every mistake, glitch, and customer complaint in your facility or department. Then calculate the cost of
finding and fixing each one. How much time, energy, and money
does that take away from doing your *real* business? What does it
cost? If you weren't fixing the mistakes, what could you be doing
instead? Multiply this by the number of days in the week, month,
or year. Ouch!

   These errors aren't your fault, and they're not the fault of your
people. It's your systems and processes that are at fault; they let
people make mistakes that could be prevented.

> **Hint: Blame your processes, not your people.**

EVERY BUSINESS PROCESS HAS THREE BIG LEAKS

It doesn't matter if you're in manufacturing or services, healthcare
or groceries, injection molding or consulting; every business is

FIGURE 6.2

Six Sigma Cost of Poor Quality

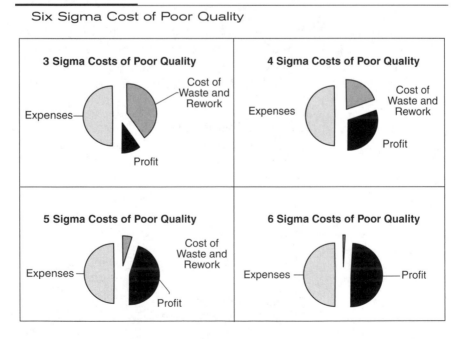

leaking cash. Most 3 sigma businesses try to blame these problems on their employees, but the problem isn't the people.

> **Big leak 1:** *Delays.* The delays between the steps in your process cost you time and money that dampen your productivity and profitability and increase the chances for error.

> **Hint: Watch your patients, not your people!**

> **Big leak 2:** *Defects.* These are the defects, mistakes, and errors that have to be fixed or scrapped. Fixing mistakes that shouldn't have been made in the first place consumes time and money that could be better spent serving customers and boosting the bottom line.

> **Big leak 3:** *Deviation.* These are the small to large differences from patient to patient, day to day, and month to month of your products and services.

> **Hint: Forget specification limits; focus on target values. Tiny deviations from your customer's target value cost time and money.**

Even a small reduction in delay, defects, and deviation in your mission-critical processes can give you a sustainable competitive advantage. Customers aren't stupid. They can tell a finely tuned hospital from a clumsy one. Once you have a head start, your competitors always will be playing catchup.

EVERY BUSINESS HAS TWO IMPROVEMENT FOCUSES

Every business consists of (1) the core business activity and (2) the supporting operational processes.

1. *The core business.* In a hospital, it's the diagnosis and treatment of patients that involve doctors, nurses, lab work, and so on. A printer focuses on getting an image on some kind of medium. A manufacturer focuses on getting products manufactured to specifications. A hotelier focuses on a customer's stay. With all the data I've looked at, even

caring for patients in a hospital, no business is better than
0.6 percent error (6,000 mistakes per million). The 1999
study, *To Err Is Human* (National Academy Press, 2000),
found that mortality rate in all aspects of hospital care
made it the eighth leading cause of death in the United
States. *An article in 2009 indicated that this hasn't improved.*

> **Hint: Even if you are 99 percent good,
> fixing the 1 percent bad can cost a fortune.**

2. *Operations*. Operations includes every aspect other than
the core business: marketing, sales, orders, purchasing,
billing, payments, etc. I've seen data that show a 3 percent
error rate on patient armbands, 17 percent order errors,
and $100 million dollars in rejected insurance claims.
These are all operational problems.

Most businesses spend too much time working on their
strengths (making the core business process more effective and efficient) and too little time working on their weaknesses (marketing,
sales, invoicing, billing, shipping, purchasing, and payments).
While the customer-affecting improvements to the core business are
important, the profit-affecting ones on the operations side are critical to reducing costs and boosting profit. To make breakthrough
improvements in speed and quality possible, you have to take some
time out of your busy schedule and shift your focus. I'm not asking
for a full-time commitment, just a couple of hours per week.

Success Secret 1: Work on Your Unit, Not in It

I recently went into Sears to order a dishwasher and a TV. I got the
part numbers and went to one of the checkouts in Appliances. The
cashier said that she could order the dishwasher but not the TV. I'd
have to go to the TV Department to order the TV. The TV
Department wanted to charge me double to have it delivered on
the same day as the dishwasher. Doesn't this sound stupid to you?
Shouldn't I have been able to order and pay for them both at the
same time?

Have you ever walked into someone else's business and almost
immediately noticed some way that the person could improve the

operation to be better, faster, or cheaper? Why hasn't the owner noticed what you find obvious?

The answer isn't obvious: The owner is busy working *in* his or her business, but he or she rarely ever steps out and works *on* his or her business.

We all get trapped mentally inside of our hospitals and our orientations because we spend so much time working *in* them. It takes some mental gymnastics to learn how to step outside the business, to get some distance from it, so that you can work *on* the business and its processes. If you want a reliable, dependable hospital that produces predictable, consistent results, you will need proven methods and tools to make it happen.

Success Secret 2: Watch Your Process, Not Your People

Startup businesses succeed because smart people figure out how to turn a profit. Customer-serving processes grow up in an ad hoc fashion. Business owners come to rely on their people, not their processes, to deliver a consistent return on investment.

Because companies often start from humble beginnings and grow rapidly beyond their grassroots capabilities, it's easy to get hooked on the excitement of crisis management and firefighting. It becomes a way of life in most businesses. When daily heroics are required to avoid missing commitments and preventing mistakes, companies come to rely on heroes. The clinical side of healthcare is especially prone to this process. There's even a place dedicated to heroics: the emergency department (ED).

This is another mistake. It often comes from your business orientation. *People-oriented* companies focus their attention on *who* is doing the job—doctors and nurses. People-oriented businesses believe that quality and productivity are a function of their people, not their processes. They think, "If I could only get the right person in this job, everything would be peachy." Unfortunately, great people come at a premium price, and when they leave, they take their wisdom and process with them. Sounds like healthcare, doesn't it?

Process-oriented businesses, on the other hand, rely on mistake proof processes to ensure that care is delivered on time and error-free. Process-oriented companies focus on developing and following the right process. They depend on good processes to produce superior results. Here's some good news: With a great process, you can hire and train the lowest-skill-level people available. They have procedures for everything from cleaning restrooms (e.g., McDonald's) to maintaining Navy jet fighters. If the Air Force can teach 18-year-olds to maintain $30 million jets, you can develop processes that anyone can follow.

Hospitals all over the nation, for example, have to deal with "codes" when a patient's vital signs crash. Fewer than 5 percent of the patients can be revived. Based on research done in Australia, most hospitals are implementing rapid-response teams (RRTs) to prevent codes. There are a few key vital signs that indicate that a patient is heading for a code; nurses are being trained to identify these trends and call in an RRT. The hospitals that have implemented RRTs have cut their codes (and mortality rates) by half or more. Similarly, hospitals have identified a few key procedures and therapies that can prevent problems such as heart attacks, heart failure, ventilator-acquired pneumonia, and infections. Some of these are as simple as giving an aspirin at arrival and discharge. The Institute for Healthcare Improvement (IHI) estimates that these therapies saved 122,346 lives over an 18-month period from 2004 to 2006. This is the power of good processes. Not only do they save time and money, but they also save lives.

When you have good processes, there's less need for overtime, and you can hire the lowest-skill-level person for the job. Labor costs are cheaper because you are not bidding for a small group of the best people; you can hire anyone and train them for the job.

Success Secret 3: Watch Your Patients, Not Your People

If you watch the doctors and nurses in your hospital, they're usually busy. Watch patients work their way through your facility, and you'll most likely find that they're only being cared for about 3 minutes out of every hour. The rest of the time they're waiting for something to happen. Look at any waiting room: Patients and family members waiting. Look at most ED exam rooms: Patients waiting.

If you want to learn how to make your unit more useful, don't bother watching your coworkers. Watch your patients. What are they doing? Maybe you can easily see ways to make healthcare more beneficial, easier to use, less likely to fail, and so on.

Success Secret 4: Watch Your Product, Not Your People

Trying to make employees more efficient is usually a waste of time; a 50 percent improvement in employee efficiency will barely make a dent in your overall cycle time Making your product or service more efficient is a great use of time. How long does it take to gather all the information to issue an invoice or bill? Why isn't it all up to date and available immediately? Why does a purchase order take so many approvals? Why does it sit in so many in-baskets waiting for a signature? Face it, your product or service is lazy. It's sitting and waiting for someone to work on it over 90 percent of the time. Watch your product, not your people.

When you take these secrets to heart and start making improvements, you'll see a rapid improvement in the bottom line.

Success Secret 5: Implement a Proven Improvement System

Because of this people orientation, most managers and employees think that they should be able to find and fix problems in their business using their instincts, and they can, up to a point where they hit a wall. This isn't their fault. Research into the science of change has found that one set of problem-solving methods (e.g., common sense and trial and error) will work for a certain class of problems, but not another. Then you will want to discover a new set of methods and tools to solve the next class of problem. Consider antibiotics: They fight bacterial infections but not viruses such as the common cold. The same is true in business.

Since most processes are created by accident in an ad hoc way, problems with the processes are fixed using *common sense* and *trial and error* as the business grows. At some point, however, the ability of these two methods to solve the more mysterious and complex problems begins to fall off. Eventually, they stop working all together. This early success and later failure syndrome affects all problem-solving methods.

Throughout time, people have routinely found ways to solve seemingly unsolvable problems. Edison invented the light bulb. The Wright brothers figured out how to fly. But, to do this, they invariably had to invent new ways to solve problems that exceeded the grasp of the old methods.

Fortunately, the methods and tools for creating and improving your processes and systems have already been developed and proven in every industry. Lean Six Sigma has a seemingly bottomless pit of tools and techniques to make improvements, but I have found that a few key tools used in the right sequence are all you need to start making immediate breakthrough improvements in speed, quality, productivity, and profitability.

Every business has to improve the key aspects of performance every year just to keep even with the competition. The only question is whether you're going to rely on the declining effectiveness of common sense and trial and error or are you going to upgrade your ability to solve the stubborn, seemingly unsolvable problems? If you aren't going to employ the proven strategies of Lean Six Sigma simplified, what are you going to do instead?

Turn your hospital into an asset that produces predictable results. Don't let your hospital run you. Aren't you tired of dealing with the seemingly unrelated problems that occur every day in the hospital? Haven't you waited long enough to find a new and improved way to plug the leaks in your cash flow?

THE UNIVERSAL IMPROVEMENT METHOD

Give a man a fish and you feed him for a day;
Teach a man to fish and you feed him for a lifetime.

—ASIAN PROVERB

Regardless of the acronym used for describing business process improvement—TQM, PDCA, DMAIC, DFSS, etc.—the overarching method is always the same. My acronym for this method is FISH—*f*ocus, *i*mprove, *s*ustain, and *h*onor. Few companies achieve success "overnight." Companies that achieve lasting success do so by getting better *over time*. They've learned the secrets of knowing how to FISH.

Life and business involve a series of incremental sustaining improvements punctuated by periodic dramatic and disruptive improvements. These breakthrough improvements or process innovations rarely can be planned but rather come about as a result of focused improvement. Invariably, this process of personal and professional evolution involves four key steps:

1. *Focus* on one key problem, skill, or area of your business life at a time.

2. *Improve* significantly in that area.

3. *Sustain* the improvement through repetition and practice until it becomes an unconscious habit. Measure and monitor to ensure that you sustain the new, higher level of performance.

4. *Honor* your progress through simple rewards. Then review what you've learned, and refocus on another area for improvement.

This simple process is the secret of mastering every aspect of your business. You won't do it overnight, but you will over time!

Focus

One arrow does not bring down two birds.

—TURKISH PROVERB

Who begins too much accomplishes little.

—GERMAN PROVERB

Most people are unclear about what they actually want from their business. This lack of clarity translates into confusion about what to do and when to do it.

The secret of success is to avoid trying to do everything and instead focus on the most important, highest-leverage things to improve. Far too many people "major in minor things," as Zig Ziglar would say.

The 4-50 Rule. Pareto's 80/20 rule states that 20 percent of what you do will produce over 80 percent of the results. In working with people and businesses, I have discovered a refinement of this rule that I call the *4/50 rule*: 4 percent of what you do will create over 50 percent of your results. This is where you should spend your time. You don't have to improve everything in your business, just a few key things that really matter.

Improve

He who would learn to fly one day must first learn to stand and walk and run and climb and dance; one cannot fly into flying.

—NIETZSCHE

Action will remove the doubt that theory cannot solve.
—TEHYI HSIEH

The only sustainable advantage may be the ability to
learn faster than your competition.
—PETER SENGE, AUTHOR OF *THE FIFTH DISCIPLINE*

Step 1: Get started, but start simply, inexpensively. Focus on one or
two broad areas: (1) eliminating delays using Lean or (2) reducing
defects or deviation using Six Sigma.

Step 2: Identify one mission-critical problem to solve. It must
be something you can affect directly. You can't, for example, fix loss
of market share directly, but you can reduce the product defects
and delays that are causing patient defections.

Step 3: Make the invisible visible. If you want to reduce delay,
defects, and deviation:

1. *Reduce delay:*
   - Flowchart or value-stream map your process.
   - Analyze where most of the delay occurs, and eliminate
     it.
2. *Reduce defects:*
   - Count your misses, mistakes, and errors, and plot them
     on a control chart. You will need control charts to moni-
     tor the improvement, so start with one.
   - Categorize your misses, and display them using a Pareto
     chart or two. Narrow the focus to the first "big bar."
   - Analyze the root causes of these mistakes using a fish-
     bone diagram and how to prevent them using a counter-
     measures matrix.
3. *Reduce deviation:* All processes produce varying results. A
   hospital admission process may take a little more or a little less time. Housekeeping staff may take a little more or
   a little less time to clean a room. A manager may take a
   varying amount of time to make a decision. Getting bids
   for purchases will take varying amounts of time. Getting
   approvals for purchases takes a widely varying period of
   time. To *reduce variation*, you will want to
   - Measure your performance in cycle time, length, width,
     weight, volume, or money.

- Use histograms and control charts to understand the variation.
- Analyze the root causes of variation, and reduce it.

Sustain

Perhaps the most difficult part of any change is sustaining the new way of thinking, being, doing, or acting. It's easy to fall back into the old rut.

Step 1: Make the invisible visible. Start using special graphs called *control charts* and *histograms* to monitor the behavior of your processes. To use them, you don't have to be a statistician; you just have to know how to read them. Control charts will tell you when something abnormal happens to your process. There are rules built into the QI Macros software that will alert you to each potentially unstable condition so that you can take action.

Step 2: Monitor and sustain the improvement. In the beginning, be patient and open to learning about how these charts will reveal the inner mysteries of how your business works. As they alert you to changes, take action to restore the new, higher level of performance.

Honor

In every work, a reward added, makes the pleasure
twice as great.

—EURIPIDES

Most businesses are constantly improving, but sometimes they forget to take time to honor their progress. There will always be more to learn and more to do. If you focus only on what you don't yet know, what you haven't yet done, you'll eventually burn out. So it makes sense, periodically, to look back over the last week, month, and year.

- What worked? What have you learned?
- What have you accomplished?
- How have you grown?
- What's next?

Life is often lived in fits and starts, moving ahead and falling back, but in general, with the right set of starting beliefs and values, the quality of life improves. Where were you 5 or 10 years ago? What has improved? What have you let go of that you no longer need? Without rewards, anyone eventually will give up their quest for improvement. And since the outside world is busy and sometimes thoughtless, you'll need to figure out how to reward and recognize the improvement teams and process.

Develop a system of rewards and recognition. Once a mind connects pleasure with improvement, you'll be surprised by the quality and quantity of ideas. Once you've identified, improved, and sustained a new level of performance in one area of your business, something else will become more vital to your personal and professional evolution.

How will you know what to focus on next? Return to your measurements. What's next?

- Delay?
- Defects?
- Deviation?

LEAN SIX SIGMA

Lean Six Sigma will focus your improvement efforts to drive dramatic improvements in speed, quality, and profitability. The methods and tools of Lean will help to drive dramatic improvements in speed and productivity. The methods and tools of Six Sigma will help to drive radical reductions in defects and deviation that will improve productivity and profitability. Regardless of the acronyms used or the number of steps, Lean Six Sigma follows a universal improvement process: Focus, improve, sustain, and honor (FISH). There is a handful of tools that you will need for each of these steps to move from 3 to 5 sigma. To rise to Six Sigma, you will need some more robust tools, but you won't be ready for their rigor until you've embraced and mastered the basic tools.

There are some additional methods and tools that you can use to design innovative products and processes from scratch. These are called *Design for Lean Six Sigma* (DfLSS or DFSS).

Change happens in projects.

—KEN MILLER

While most books start you on the path toward total domination of the corporate culture and business processes, I'd like you to start by piloting some focused improvement projects involving Lean and Six Sigma. As you begin to master the improvement processes, then, and only then, would I like you to consider expanding the scope to include more people and projects to the point that Lean Six Sigma becomes a way of doing business, not just a program of the month or the pet project of a CEO.

The methods and tools are the easy part; *changing culture is hard*. When you start by creating successful projects and let the corporate grapevine sell Lean Six Sigma for you, it will be easy to change the culture because the culture will adopt and adapt Lean Six Sigma on its own. When you start by trying to force Lean Six Sigma down everybody's throat with endless training, changing the culture can get hard, if not impossible.

Lean Six Sigma will not fix everything about your business. It won't fix suppliers. It won't fix customers. It won't fix morale. It won't fix poor leadership. But it is a management system that can improve morale, leadership, and patient satisfaction indirectly. Learning Lean Six Sigma will help you to choose and improve your suppliers. It will help you to understand and better serve your existing and undiscovered customers.

Take some time to test drive each of the improvement methods and tools. Apply them to your business and your processes. Use the QI Macros tools to focus, improve, sustain, and honor your progress. You'll be surprised how easy it can be to find and make dramatic improvements. Best of all, these methods and tools have stood the test of time. You'll be able to use them in any business and any job you ever have. And you will be recognized because you're the employee who can find the hidden gold mine in the business.

What could you accomplish with Lean Six Sigma?

CHAPTER 7

Excel Power Tools for Lean Six Sigma

While Lean doesn't require many tools other than a pad of Post-it Notes, Six Sigma thrives on charts, graphs, and diagrams of performance data. To succeed at Six Sigma, you'll need a set of power tools.

When I learned quality improvement back in 1989, I had to draw all the Six Sigma charts by hand. I spent five days in a control chart class calculating all the formulas using a handheld calculator and drawing the various charts. Most of my fellow students and I struggled to calculate the formulas correctly. At the end of the course, the class spent only two hours on what the charts were telling us. I knew there was no way I was going to get phone company personnel to draw control charts by hand, but I couldn't get my management to spend $1,000 on statistical process control (SPC) software, so I just struggled along.

After I left the phone company in 1995, I started experimenting with using Excel to draw all the charts necessary for Lean Six Sigma. I launched the first, primitive version of the QI Macros in 1997 and have been improving them ever since. Since Excel does all the heavy lifting—drawing the charts—I can keep the cost low enough for the typical user.

Microsoft Excel is a tremendously powerful tool for Lean Six Sigma, but most people don't even know how to use the basic capabilities of Excel. If you think you're a hotshot Excel user, read on because we'll look at how to use the QI Macros Lean Six Sigma SPC Software for Excel. If you're not that familiar with Excel and how to set up your data to make them easy to analyze, chart, and graph, then you will get a lot from this discussion. If you don't own a copy of Excel or Office, you usually can pick up inexpensive copies of older versions at ebay.com. The QI Macros work in all versions of Excel.

THE QI MACROS FOR EXCEL

The QI Macros Lean Six Sigma SPC Software consists of five main parts:

- Over 30 tools to draw control charts, histograms, Pareto charts, and so on. The Control Chart Wizard will automatically choose the right control chart for you.
- Over 90 fill-in-the-blank templates of Lean Six Sigma forms, tools such as the fishbone diagram, and charts such as the XmR control chart.
- Two fill-in-the-blank dashboard tools for XmR, c, np, p, and u charts.
- Statistical tools such as analysis of variance (ANOVA), *t* tests, regressions, and so on.
- Data transformation tools such as the PivotTable Wizard, word count, stack, and restack.

THE QI MACROS ARE EASY TO USE

Since I'd never been exposed to SPC software developed before the now-familiar point-and-click, mouse-driven interface, I was free to think outside the interface design imposed by minicomputers. I took a "grab it and go" approach to the software—select data with the mouse, and then click on a menu to draw a chart. The QI Macros are easy to learn and use:

- Because the software was developed from the ground up to work in Excel and deliver immediate results in business environments using "grab it and go" simplicity:
  - *Mistake-proof selection of data.* Your data can be in connected or separated rows or columns; the QI Macros will clean up nonnumeric data, fix any misalignments, and use your data as you selected them.
  - *Control Chart Wizard* to select the right control chart for you automatically.

- *Control Chart Dashboards* to simplify monthly reporting.
- *PivotTable Wizard* to simplify analyzing complex transaction files.
- *Mistake-proof statistical analysis.* Excel can be picky and even produce invalid results if the data aren't used correctly.

- Because the QI Macros do all the math and statistics for you. There are no complex formulas to grasp or apply, just charts and results.
- Because of the fill-in-the-blanks, paint-by-numbers simplicity of the 90+ chart and documentation templates.
- Because you don't have to waste time transposing or transferring your data from Excel to a separate program. The QI Macros work inside Excel.

QI MACROS INTRODUCTION

There are many graphs, forms, and tools used in Lean Six Sigma and SPC. There are four key elements of the QI Macros: macros, templates, data transformation, and statistics.

Ninety percent of common problems can be diagnosed with control charts, histograms, Pareto charts, and Ishikawa diagrams. A couple of control charts will help you to sustain the improvements. Microsoft Excel can be used to create all these charts, graphs, forms, and tools.

Installing the QI Macros

To install the QI Macros, simply

1. Go to our Web site www.qimacros.com/hospitalbook .html, and fill in your e-mail address to download the QI Macros and the other free Lean Six Sigma quick reference cards. This also will sign you up for the free QI Macros and Lean Six Sigma lessons online.

2. Download the QI Macros 90-day trial copy by clicking on the CD icon.

3. Double click on QIMacros90day.exe to install the QI Macros.

4. When you start Excel, the QI Macros menu will appear on Excel's toolbar in Excel 2000–2003 or the ribbon menu in Excel 2007–2010.

5. If you have any problems, check my Web site:
www.qimacros.com/techsupport.html.

Sample Test Data

The QI Macros for Excel installs test data on your PC in My
Documents/QI Macros Test Data. Use these data to practice with
the charts and to determine the best way to format the data before
you run a macro.

Creating a Chart Using the QI Macros Menu

There are two different ways to create charts in the QI Macros:

1. Select your data and then run a macro from the menu. To
   create a chart using a macro from the menu, just select the
   data to graph using the mouse. Then, using the QI Macros
   menu (Figure 7.1), select the chart you want to create. The
   QI Macros will do the math and draw the graph for you.
2. Use the fill-in-the-blanks chart templates.

Create a Control Chart

1. *Open a workbook* (e.g., Healthcare SPC.xls, which has
   worksheets with healthcare data for the common control
   charts and Pareto charts).
2. *Select the labels and data to be graphed* (e.g., XmR data—Falls
   per 1,000 patient-days). Click on the top left cell, and drag
   the mouse across and down to include the cells on the
   right.
3. *From the QI Macro menu, select Control Chart Wizard.* Excel
   will start drawing the graph. Fill in the graph title, and
   the x and y axis titles as appropriate.

FIGURE 7.1

QI Macros menu.

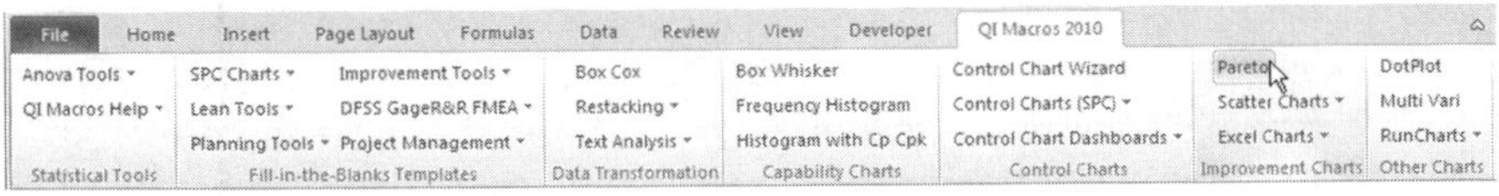

4. *To add text to any part of the graph, just click anywhere on the white space and type.* Then use the mouse to click and drag the text to the desired location. To change titles or labels, just click and change them. Change other text in the worksheet in the same way.

5. *To change the scale on any axis, double click on the axis.* Select "Scale," and enter the new minimum, maximum, and tickmark increments.

6. *To change the color on any part of the graph, double click on the item to be changed.* A patterns window will appear (Figure 7.2). Select "Font" to change text colors, "Line" to change line colors and patterns, or "Marker" to change foreground and background colors. Line graphs showing defects or delay are the key first step of any problem solution.

7. *To change the style of any line on the graph, double click on the line.* The format line window (Figure 7.3) is displayed. Changing the line style, color, and weight is all performed in this window. When you're done, click "OK." The changed graph is now easier to read.

F I G U R E 7.2

Excel chart patterns window.

FIGURE 7.3

Excel format line window.

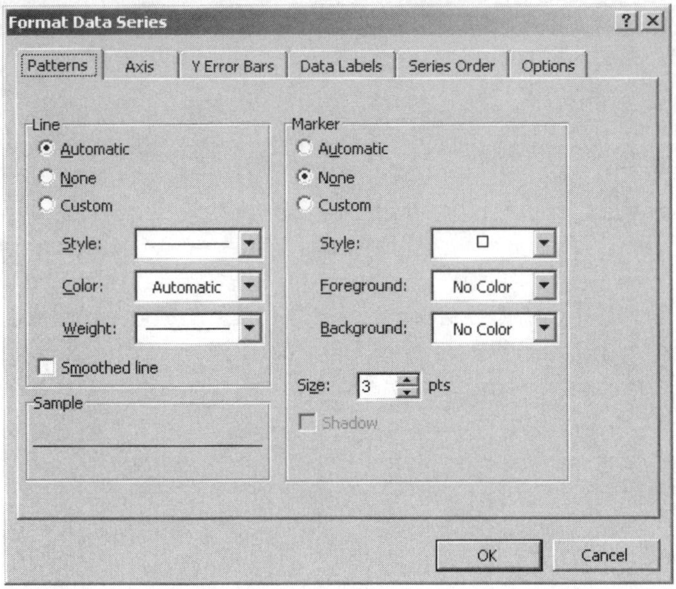

8. *To change the style of graph, right click on the chart and choose "Chart Type."* Click on the desired graph format, and then click "OK."

Fill-in-the-Blanks Templates

In addition to the charts listed on the menu, the QI Macros contain over 80 fill-in-the-blank templates. To access these templates, select the "Fill-in-the-Blanks Templates" on the QI Macros menu (Figure 7.4).

FIGURE 7.4

QI Macros Fill-in-the-Blanks Templates menu.

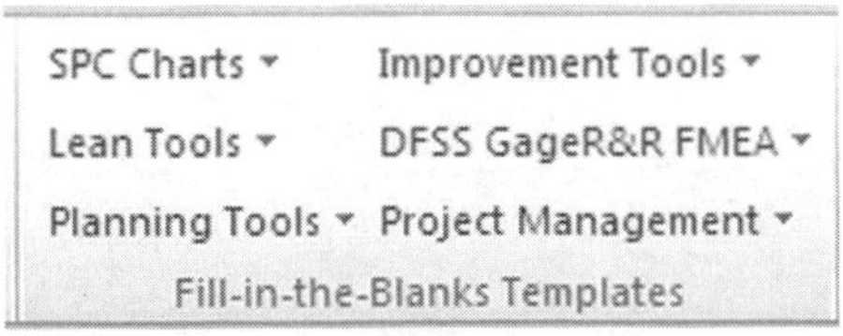

Each template is designed for ease of use. Tools such as the flow-chart and fishbone diagram make use of Excel's drawing toolbar. To view Excel's Drawing Toolbar, select "View—Toolbars-Drawing" in Excel 2000–2003 or "Insert—Shapes" in Excel 2007–2010.

Creating a Control Chart with a Template. To monitor and control performance, hospitals will want to add data to the charts every month. To simplify this process, the QI Macros contain templates for each kind of control chart. Just cut and paste, or input data directly into the yellow area. The control charts will populate as you input the data.

These templates are especially helpful if you have novice personnel (e.g., at nursing stations) who will be inputting data or you don't have enough data to run a macro (you're just starting to collect the data). To create a chart using a template, click on the QI Macros menu and select the "Fill-in-the-Blanks Templates" (Figure 7.5). Click on the template you want to use (e.g., "SPC Charts—g chart").

The input areas for most of the templates start in column A (see Figure 7.5). Either input your data directly into the yellow cells on the template, or cut and paste the data from another Excel spreadsheet. As you input the data, the chart will populate to the

FIGURE 7.5

QI Macros g chart template.

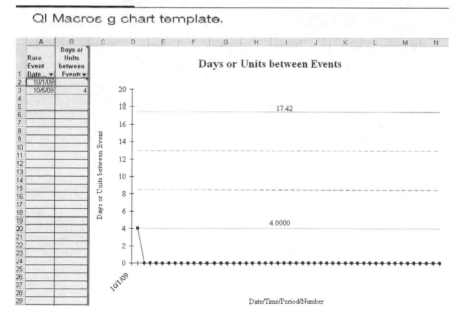

right. The X chart templates also display a histogram, probability plot, and scatter plot.

Running Stability Analysis on a Chart Created by a Template. To run stability analysis on a chart created using a control chart template, click on the chart (dark boxes will appear at the corners), click on the QI Macros menu (Figure 7.6), and select "Analyze Stability."

Choosing Which Points to Plot. Each template defaults to 50 data points. If you have fewer than 50 data points and only want to show the points with data, click on the arrow in cell B1. This will bring up a menu (Figure 7.7). Select nonblanks to plot only the points with data.

In addition to control charts, there are templates for histograms with Cp and Cpk, precontrol charts, probability plots, Pareto charts, and many more.

The control chart dashboards have a data worksheet for up to 120 measurements. Once the data are entered, a single click of the "Create Dashboard" button will create a single sheet listing all the charts. Each month, just add data to the worksheet, and press "Update Charts" to update all the charts.

Templates for Your Quality-Improvement Efforts. Examples of other templates you will find in the QI Macros are as follows:

- *Focus your improvement efforts* using the balanced scorecard, tree diagram, voice-of-the-customer matrix, or cost-of-quality template.

F I G U R E 7.6

QI Macros chart menu.

FIGURE 7.7

Eliminating blanks from QI Macros templates.

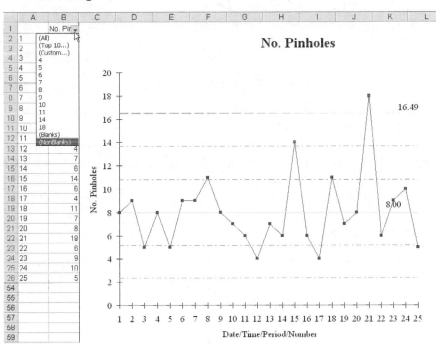

- *Reduce defects* using the Pareto chart, Ishikawa or fishbone diagram, and countermeasures matrix.
- *Reduce delay* using the value-stream map, flowchart, value-added flow analysis, time tracking, and takt time templates.
- *Reduce variation* using the control charts and histograms.
- *Reduce measurement error* using the gauge R&R template.
- *Design for Lean Six Sigma* using the failure modes and effects analysis (FMEA), QFD house of quality, Pugh concept selection matrix, and design of experiments.
- *Engage in project management and planning* using the Gantt chart, action plan, and return-on-investment (ROI) calculator.

Put Your Whole QI Story in One Workbook

Because the QI Macros are an all-in-one toolkit for Lean Six Sigma, you can put your entire improvement story in one workbook simply by adding worksheets. Let's say that you've created a control chart and a Pareto chart in one workbook. After you choose "Ishikawa/Fishbone" from the "Fill-in-the-Blanks Templates," just right click on the sheet name, and select "Move" or "Copy Sheet" to move the template into the existing workbook. It's a great way to keep all your information in one place.

Data Transformation

Convert Tables of Data from One Size to Another. Sometimes data have to be reorganized or summarized before they can be graphed. What do you do when your gauge or database gives you a single column of data that actually represents several samples (Figure 7.8)? How do you convert it to work with the XbarR or other chart?

FIGURE 7.8

Single column of data.

	A	B
1	**Drug**	**Diffrate**
2	Drug 1	8
3	Drug 1	4
4	Drug 1	0
5	Drug 1	14
6	Drug 1	10
7	Drug 1	6
8	Drug 2	10
9	Drug 2	8
10	Drug 2	6
11	Drug 2	4
12	Drug 2	2
13	Drug 2	0
14	Drug 3	8
15	Drug 3	6
16	Drug 3	4
17	Drug 3	15
18	Drug 3	12
19	Drug 3	9

1. Select the single column of data.
2. Click on "Data Transformation—Stack/Restack" to choose various tools, including a restack matrix.
3. Enter the number of columns (e.g., 5), and click "OK." The macro will reformat your data to five columns and however many rows. For example, if you have 18 data points and you input 6 into the prompt, you will get six columns and three rows of data.

Summarize Your Data with Pivot Tables. The QI Macros will draw graphs, but they won't summarize your data because they cannot read your mind. However, you can use Excel's PivotTable function to summarize data in almost any conceivable way. For example, what if you have a series of report codes from a computer system or machine? You need to summarize them before you chart them. Just select the raw data, and go to Excel's menu bar and choose "Data—PivotTable." With a little tinkering, you'll learn how to summarize your data any way you want them.

With the QI Macros, it's easy to create a pivot table. Just use the mouse (and the Alt key) to select up to four headings in the sheet. Then run "Data Transformation—PivotTable Wizard." The wizard will guess how best to organize the data selected into a pivot table.

Or, using Excel, you can do it the manual way:

1. *Select the labels and data to be summarized* (Figure 7.9), in this case, denied charges by date, facility, and region. Many computer systems produce one code or measurement each time an event happens. These often need to be summarized to simplify your analysis.
2. *Using the QI Macros, click on up to four column headings, and choose "Data Transformation—PivotTable Wizard"* (Figure 7.10).
3. *In native Excel, choose "Data—Pivot Table" (Excel 1997–2003) or "Insert PivotTable" (Excel 2007–2010).* Click "Finish" to get a screen like the one in Figure 7.11.
4. *Click, hold, and drag the data labels into the appropriate area of the pivot table to get the summarization you want* (see Figure 7.10).
   - *Page fields:* Use this for big categories (e.g., vendor codes, facilities in a company).

FIGURE 7.9

Pivot table data.

	A	B	C	D	E	F	G	H	I	J	K
1	Region	POST DATE	ENT	ADM DATE	DIS DATE	AS	COS	FC	IN1	PT	DENIED CHARGES
2	North	6/27/03	Hosp1	2/13/03	1/1/00	OL		X	AEH	O	543.07
3	South	12/24/02	Hosp2	7/13/02	1/1/00	OL		X	BCP	E	215.4
4	South	2/25/03	Hosp2	12/6/02	1/1/00			X	CGH	O	157.92
5	South	5/23/03	Hosp3	10/20/02	1/1/00	OL		X	MAH	O	90.73
6	North	7/15/03	Hosp1	5/7/03	1/1/00	AP		X	HEH	O	4103.78
7	North	11/5/02	Hosp4	8/6/01	1/1/00	OL		F	PTB	E	3224.83
8	North	11/20/02	Hosp5	4/15/02	1/1/00	OL		F	PTB	O	3291.76
9	North	11/27/02	Hosp1	5/13/02	1/1/00	OL		F	PTB	O	13845.9
10	North	11/27/02	Hosp4	9/16/02	1/1/00			F	PTB	O	1151

FIGURE 7.10

Pivot table results.

	A	B	C	D	E	F	G	H	I
1	Region	(All)							
2									
3	Sum of DENIED CHARGES	ENT							
4	ADM DATE	Hosp1	Hosp2	Hosp3	Hosp4	Hosp5	Hosp6	Hosp7	Grand Total
5	3/28/00			387.48					387.48
6	4/25/00			379.62					379.62
7	3/13/01			6908.98					6908.98
8	7/24/01		311.16						311.16
9	7/26/01					2124.86			2124.86
10	8/6/01				3224.83				3224.83
11	8/20/01		193.65	343.51					537.16
12	10/23/01			230.42					230.42
13	11/16/01			2186.16					2186.16
14	11/19/01			2627.84					2627.84
15	11/26/01			311.2					311.2

- *Left column:* Use this to summarize by dates or categories.
- *Top row:* Summarize by subcategories.
- *Center:* Drop fields to be counted, summed, or averaged into the center.

4. *To change how the data are summarized, use the PivotTable Wizard or double click on the top left-hand cell.* For online tutorials, Google "Excel Pivot Table."

5. *Select labels and totals, and draw charts using your summarized data.*

FIGURE 7.11

Pivot table layout window.

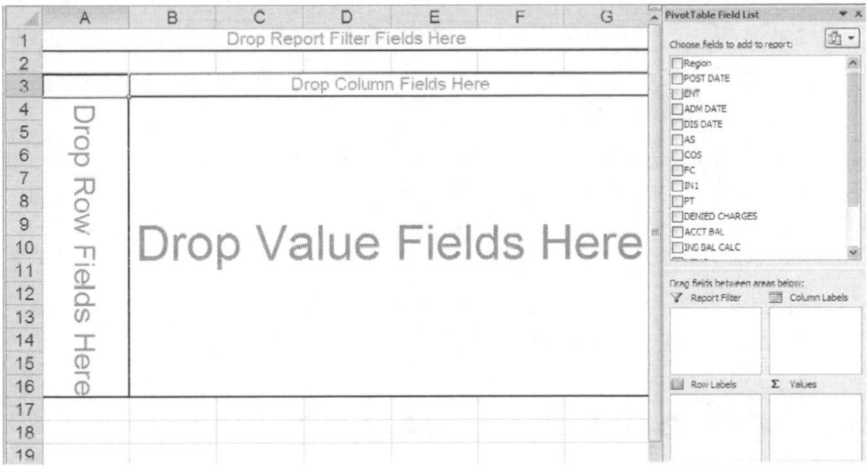

Using the ANOVA and Other Statistical Tools

Most Six Sigma black belts get into more detailed analysis of data to determine the variation. Analysis of variance (ANOVA) seeks to understand how data are distributed around a mean or average.

To perform ANOVA in native Excel, you must have Excel's Data Analysis ToolPak installed. Go to "Tools—Addins" (Excel 1997–2003) or "Excel Options—Addins—Manage Excel Addins," and check "Analysis ToolPak" (Figure 7.12). Excel will either turn

FIGURE 7.12

Turning on the Analysis ToolPak.

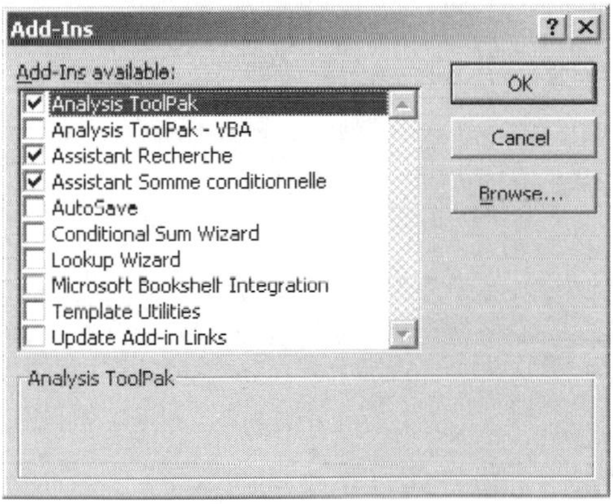

these tools on or ask you to install them using your Office or Excel CDs. To check if they have been installed, click on "Tools—Data Analysis." If you cannot see "Data Analysis" in the Tools menu, the statistical analysis tools are not installed.

1. Select the data to analyze. These data must be organized in columns.
2. From the QI Macros pull-down menu, select "ANOVA and Other Analysis Tools."
3. Click on the appropriate analysis tool (ANOVA, regression, f test, t test, etc.).
4. See sample test data for each tool, and test on your computer at c:\qimacros\testdata.

POWER TOOLS FOR LEAN SIX SIGMA

As you can see from these examples, Excel and the QI Macros are power tools to simplify Lean Six Sigma. By putting your data into Excel, summarizing them with pivot tables, and graphing them with the QI Macros, you can automate and accelerate your journey toward Six Sigma.

1. The QI Macros give you the power to select data and immediately draw all the key charts and diagrams: Pareto charts and fishbone diagrams for problem solving as well as histograms and control charts for reducing variation.
2. The QI Macros templates give you fill-in-the-blanks simplicity for control charts, Pareto charts, fishbone diagrams, flowcharts, and value-stream mapping.
3. The QI Macros ANOVA and analysis tools give you simplified access to Excel's statistical tools and much more.
4. Control chart dashboards simplify monthly reporting.
5. Data transformation tools help to reorganize your data or put them into pivot tables.

Start using Excel and the QI Macros to organize, analyze, and graph your data to illuminate the opportunities for improvement.

ANALYZING CUSTOMER SERVICE DATA HIDDEN IN TROUBLE-REPORTING SYSTEMS

In service industries, much of the information you need to make breakthrough improvements is buried in trouble-reporting systems. Call-center personnel routinely attempt to capture customer complaints, categorize them, and include remarks about the customer's dilemma. Unfortunately, the categories in most information systems are predefined, inflexible, and rarely speak to the true nature of the customer's complaint. And often the customer, who has waited in a call queue for several minutes, has had time to think up several questions that he or she needs answered, not just one.

In these situations, the information needed to analyze these customer interactions is in the freeform remarks, not in the convenient categories. The information captured in the remarks invariably will be more accurate than that in the predefined categories. How do we analyze this wild potpourri of short phrases and abbreviations? The answer lies in Microsoft Excel.

Importing Text with Microsoft Excel

To analyze text with Excel, you must first import the data into Excel. To do this, you will need to export the customer account and remarks information from the trouble-reporting system into your PC or local-area network.

To simplify deeper analysis, it will be useful to have some information about the customer's account included with the remarks. In a phone company, for example, having the customer's phone number will enable further analysis by allowing you to dig into the customer's records.

Then go to Excel and choose "File—Open," select "Files of Type: All Files," and open the text file. Excel's import wizard then will guide you through importing the data. Text data can either be *delimited*, which means that it contains tab, comma, or other characters that separate fields, or *fixed width*, which means that the data are of a consistent length.

The maximum number of characters Excel will store in a cell is 255, so longer text fields should be edited to fit. More than one cell can be used to store an entire remark or comment. Excel will allow up to about 65,000 rows to be imported per Excel worksheet.

Analyzing Text with Excel

Call centers often collect vital information necessary for improvement. Searching the call-center comments in an imported text file couldn't be easier. In the QI Macros, select the cells filled with text (Figure 7.13), and choose "Data Transformation—Word Count," and the QI Macros will parse the sentences into one- and two-word phrases and then pivot table them to create a list of the most common words or phrases (Figure 7.14). This often identifies the most common issue even when service representatives are using different words, phrases, or acronyms. For sample data, look in the QI Macros Test Data for crosstab.xls, and click on the "Word Count" sheet.

In this example, the most common phrase for rejected claims is "Dup DOS," for *duplicate date of service*. Notice how many ways reps phrase this: "visits same," "multiple visits," and "dos multiple." This means that we have a problem with duplicate date of service, and it's a project worth fixing.

F I G U R E   7.13

Call-center comments.

	A
1	**MEMOTEXT - Wordcount**
2	DA REJECT - DUP DOS (MULTIPLE VISITS SAME DAY). CLD AND TT SHELLY AT AETNAWAS TOLD THAT THIS SHOULD BE INCLUDED IN THE SURGERY CHARGES. E-MAILED GARY S TO SEE IF THIS IS CORRECT. THE SURGERY ACCT IS IN MC STATUS. CATHY X266
3	DANIEL FROM DAVIS WIRE 0 CALLED TOSAY THEY RECVD A W/C CLAIM FOR THIS PT AND THEY ARE NOT ASSOCIATED WITH DAVIS WIRE IN THAT REGARD...THE INS INFO IS INCORRECT...WILL DELETE AND REFER TO WC FOLLOWUP...ADAVIS
4	DA REJECT - DUP DOS (MULTIPLE VISITS SAME DAY).
5	DA REJECT - DUP DOS (MULTIPLE VISITS SAME DAY)/ RECEIVED FROM ,AILHANDLERS EOB THAT THEY WILL NOT PAY CLAIM/ CALLED TT KATHY/ SHE STATED CLAIM NOT PAID BECAUSE THEY RECEIVCED 2CLAIMS FROM US WITH SAME DOS/ NON WERE MARKED CORR
6	DA REJECT - DUP DOS (MULTIPLE VISITS SAME DAY).RCVD EMAIL BACK FROM ED THERE WAS 2 ORDERS PLACE FOR CTS VISIT 002 AT 1607 AND THE SECOND ONE FOR 003 WAS WHILE PT WAS INPT AT 0509 SENDING ACCT TO APPEALS DEPT FOR APPEAL ERIC
7	DA REJECT - DUP DOS. REQ FROM FLO THE UP SO I CAN REKEY AND CALL FOR DUP OVERRIDE. CLAIM HAS SEVERALDUPS. SANDI 26029
8	DA REJECT - DUP DOS. MEDCR DENIED FOR DUP. REQ FROM FLO A UB SO I CAN REKEY AND CALL FOR A DUP OVERRIDE. SANDI 26029
9	DA REJECT - DUP DOS (MULTIPLE VISITS SAME DAY). DELETED FRM MUO AS AUDIT REQ'D REBILL. CSCHMIDT 22984
10	DA REJECT - DUP DOS (MULTIPLE VISITS SAME DAY). DELTED CLM THAT WAS DENIED IN OCTOBER OFF MUO SUSPENCE. CSCHMIDT 22984

FIGURE 7.14

Word count of comments.

	A	B	C	D	E
1	Count of Word			Count of Two-Word Phrases	
2	Word	Total		Two-Word Phrases	Total
3	dup	11		reject dup	8
4	dos	9		dup dos	8
5	reject	8		da reject	8
6	da	8		visits same	6
7	same	7		multiple visits	6
8	visits	6		dos multiple	6
9	multiple	6		same day	5

Using Excel's COUNTIF Function

Native Excel has a function called COUNTIF that tallies cells if they match certain criteria. The formula for the COUNTIF function is

```
=COUNTIF(CellRange, "criteria")
```

The CellRange specifies the range of cells to be counted. If there is only a single column of imported text, this might be $A$3:$A$2154. Or it could include multiple columns if the text fields are longer than 255: $A$3:$C$2154.

Once you've specified the range, the real trick is to create criteria consisting of keywords and phrases that match the cells. To do this, you'll need to use Excel's *wildcard* character, the asterisk (*). To match a cell that contains a keyword, the criteria portion of the COUNTIF statement will need to look for any leading stream of characters (*), the keyword, and any trailing stream of characters (*). The simple way of expressing this in the COUNTIF statement would be

```
=COUNTIF(CellRange, "=*keyword*")
```

To make this easy to change, we might consider putting the keyword in one cell by itself and including it into the *formula*. The formula would be

```
=COUNTIF($A$1:$A$2154,"=*"&B1&"*")
```

This would take the keyword from the cell above it, making it easier to change and test various keywords. Getting the keyword right can make the resulting data more accurate.

Graphing the Data

Once you've mined all the data out of the comments, you then can use Pareto charts to examine the frequency of certain types of customer complaints. Additional digging into specific customer records may be required to determine the root cause of why these calls are being generated and how to mistake-proof the process to prevent them.

SETTING UP YOUR DATA IN EXCEL

How you set up the data can make analysis easy or hard. Using an Excel *worksheet*, you can create the labels and data points for any chart—control chart, Pareto chart, histogram, or scatter chart. This gives you a worksheet that looks like Figure 7.15.

Prepare Your Data

Data Format.   Other Lean Six Sigma software packages make you transfer your Excel data into special tables, but not the QI Macros. Just put your data in a standard Excel worksheet. The simplest format for your data is usually one column of labels and one

FIGURE 7.15

Formatting data in Excel.

	A	B	C	D	E
		Falls/1000		Total Patient	Total Patient
1	Month	Patient Days		Falls	Days
2	Jan-04	3.6		17	4658
3	Feb-04	4.5		22	4909
4	Mar-04	4.7		23	4886
5	Apr-04	6.0		30	4970
6	May-04	4.6		22	4780
7	Jun-04	3.6		18	4973
8	Jul-04	7.6		44	5762
9	Aug-04	7.7		42	5441
10	Sep-04	5.6		33	5893

or more columns of data, but the data also can be in rows. (Once you've installed the QI Macros, see c:\qimacros\testdata for sample data for each chart.)

Once you have your data in the spreadsheet, you will want to select those data to be able to create a chart. Using your mouse, just highlight (i.e., select by clicking the mouse button and dragging it up or down) the data to be graphed, run the appropriate macro, and Excel will do the math and draw the graph.

Tips for Selecting Your Data

- *Click and drag with the mouse to select the data.*
- *To highlight cells from different columns* (Figure 7.16), *click on the top left cell and drag the mouse down to include the cells in the first row or column.* Then hold down the Control key while clicking and highlighting the additional rows or columns.
- *You also may use data in horizontal rows* (Figure 7.17), *but it's not a good format for data in Excel.* While most people tend to put their data in horizontal columns to mimic the format of a calendar, this makes it difficult to use all of Excel's analysis tools. Whenever possible, put your data in columns, not rows.

FIGURE 7.16

Selecting nonadjacent cells.

	A	B	C	D	E
1	Month	Falls/1000 Patient Days		Total Patient Falls	Total Patient Days
2	Jan-04	3.6		17	4658
3	Feb-04	4.5		22	4909
4	Mar-04	4.7		23	4886
5	Apr-04	6.0		30	4970
6	May-04	4.6		22	4780
7	Jun-04	3.6		18	4973
8	Jul-04	7.6		44	5762
9	Aug-04	7.7		42	5441
10	Sep-04	5.6		33	5893

FIGURE 7.17

Horizontal data.

	A	B	C	D	E	F	G	H	I	J	K	L	M
1	Month	J	F	M	A	M	J	J	A	S	O	N	D
2	Patient Satisfaction %	82	79	84	82	92	80	94	78	83	84	92	84

- *Numeric data and decimal precision.* Excel formats most numbers as "General," not "Number." If you do not specify the format for your data, Excel will choose one for you. To get the desired precision, select your data with the mouse, choose "Format—Cells—Number," and specify the number of decimals.
- Don't select the entire column (65,000+ data points) or row (255 data points), just the cells that contain the data and associated labels you want to graph.
- When you select the data you want to graph, you can select the associated labels as well (e.g., Jan, Feb, Mar). The QI Macros usually will use the labels to create part of your chart (e.g., title, axis name, or legend). Make sure that you follow these rules when inputting your data. Make sure that you select only one row and one column of labels. Otherwise, the QI Macros will try to treat each additional row as numbers.

People often put headings for a single column into multiple cells. If you put the heading in a single cell, right click on that cell, choose "Format—Cells," click on "Alignment," and click the "Wrap Text" button. Excel will word wrap the text for you.

People also *merge* cells to have a heading span columns. Merged cells complicate *everything* in Excel. A simpler solution? Use "Format—Cells," and *center across selection.*

- *Labels should be formatted as text.* If your labels are numbers (e.g., 1, 2, 3), you need to make them text so that Excel doesn't treat them as part of your data. To do this, you will need to put text in front of them. Some examples are Sample1, S1, Lot1, and L1. If you just want the 1 to show, then you will need to put an apostrophe in front of each number to change it from data to text (e.g., '1, '2, '3, etc.). *Data should be formatted as numbers.* Your data must be numeric and for-matted as a number for the macros to perform the necessary calculations.

- *Select the right number of columns.* Each chart requires a cer-tain number of columns of data to run properly. They are
  - *One column:* Pareto, pie, c chart, np chart, XmR chart
  - *One or more columns:* line, run, bar, histogram
  - *Two columns:* scatter, u chart, p chart
  - *Two or more columns:* box and whisker, multivari, XbarR, and XbarS
- *Beware of hidden rows or columns.* If you select columns A:F, but B and C are hidden, the QI Macros will use all five columns including the hidden ones. To select nonadjacent columns, use the Control key.

DATA COLLECTION AND MEASUREMENT FOR SIX SIGMA

Six Sigma's Define, Measure, Analyze, Improve and Control (DMAIC) has an early step for measurement. While most compa-nies have too much data, people always can identify something they aren't tracking that they should be tracking. Then they think that they have to set up a whole system to collect the measurement. This is a mistake. You don't know if the measurement is useful until you have collected it for awhile. Rather than wait for a mea-surement system, start today using a few simple tools—a check sheet or a log of errors.

I use these kinds of check sheets when I'm working with a team on the "dirty 30 process" for Six Sigma Software. They find causes, and I write them down and tally the number of times each occurs. By the thirtieth data point, a Pareto pattern appears that points us to the most common (i.e., root) cause of the problem.

Check-Sheet Data Collection

Nothing could be simpler than data collection with a check sheet. The QI Macros have a template in the improvement tools to get you started (Figure 7.18). Simply print it out, and start writing on it. What to write?

In column A, write the first instance of any defect, problem, or symptom you detect. For example, if someone is calling us for sup-

FIGURE 7.18

QI Macros check sheet.

	A	B	C	D	E	F	G	H
1					Week			
2	Defect/ Problem/ Symptom	M	Tu	W	Th	F	Sa	Total
3	Delay							0
4	Missed Commitments							0
5								0
6	Defects							0
7	Errors							0
8								0
9	Repeat Fixes							0
10								0
11	Total	0	0	0	0	0	0	0

port and they have a problem with understanding a p chart, then we'd write "p chart" in A3 and put a stroke tally in the day of the week (e.g., Monday). Then we continue adding to the check sheet as the week goes on, adding defects, problems, or symptoms. By the end of the week, we'll have an interesting picture of support calls (Figure 7.19).

Just add up the number of calls, and one bucket or another will jump out as the majority of the calls. Hint: Use a Pareto chart (Figure 7.20) to show most common support calls.

Get the idea? Use a check sheet to prototype your data-collection efforts. Iterate until you start to understand what you really need to know to make improvements. A series of check sheets may

FIGURE 7.19

Check sheet of rejected claims.

	A	B	C	D
1				
2	Defect/ Problem/ Symptom	M	Tu	W
3	Duplicate Date of Service	⊪⊪ |||	⊪⊪ ⊪⊪		⊪⊪ ⊪⊪
4	No Auth	|||	||	⊪⊪

FIGURE 7.20

Pareto chart of rejected claims.

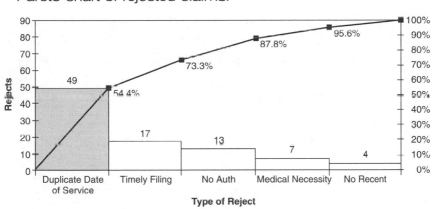

be all you need to solve a pressing problem. If necessary, you can implement a measurement system to collect the data over time.

So please don't wait for a magical, all-encompassing measurement system to deliver data. It's not going to happen. And I often find that this point of view is just an excuse to avoid making improvements ("I can't because I don't have the measurements I need").

Haven't you waited long enough to start making measurable improvements (even if your data-collection tool is just a simple check sheet)? Or are you going to keep letting loads of cash slip through your fingers. All it takes is a check sheet and a pencil. Get on with it.

Error-Log Data Collection

Another way to collect data uses an error log. In Excel, simply open a new workbook and enter headings for each data category (Figure 7.21). In this case, it's denied charges in a hospital system. Then add a new line for each denied charge. It won't take long for a pattern to emerge [payment denied owing to duplicate day of service (DOS)]. An improvement team should be able to solve this problem easily.

Sometimes these error logs get a bit more complex. To get useful data, you will want to mistake-proof the data collection.

FIGURE 7.21

Error log of denied claims.

	A	B	C	D	E	F	G	H	I	J	K	L	M	N	
1	Region	POST DATE	ENT	ADM DATE	DIS DATE	AS	COS	FC	IN1	PT	DENIED CHARGES	ACCT BAL	INS BAL CALC	MEMO	DESC
2	North	6/27/03	Hosp1	2/13/03	1/1/00	OL		X	AEH	O	543.07	543.07		DA	REJECT - DUP DOS
3	South	12/24/02	Hosp2	7/13/02	1/1/00	OL		X	BCP	E	215.4	215.4	215.4	DA	REJECT - DUP DOS
4	South	2/25/03	Hosp2	12/6/02	1/1/00			X	CGH	O	157.92	157.92	157.92	DA	REJECT - DUP DOS
5	South	5/23/03	Hosp3	10/20/02	1/1/00	OL		X	MAH	O	90.73	55.83	55.83	DA	REJECT - DUP DOS
6	North	7/15/03	Hosp1	5/7/03	1/1/00	AP		X	HEH	O	4103.78	4103.78	4103.78	DA	REJECT - DUP DOS
7	North	11/5/02	Hosp4	8/6/01	1/1/00	OL		F	PTB	E	3224.83	3224.83	3224.83	DA	REJECT - DUP DOS
8	North	11/20/02	Hosp5	4/15/02	1/1/00	OL		F	PTB	O	3291.76	3291.76	3291.76	DA	REJECT - DUP DOS
9	North	11/27/02	Hosp1	5/13/02	1/1/00	OL		F	PTB	O	13845.9	13845.9	13845.9	DA	REJECT - DUP DOS
10	North	11/27/02	Hosp4	9/16/02	1/1/00			F	PTB	O	1151	1151	1151	DA	REJECT - DUP DOS
11	North	12/11/02	Hosp6	9/24/02	1/1/00			F	PTB	O	797	797	45.38	DA	REJECT - DUP DOS
12	North	6/18/03	Hosp7	4/15/03	1/1/00			C	OT1	O	215	215	215	DA	REJECT - DUP DOS
13	North	6/25/03	Hosp5	7/26/01	1/1/00	OL		F	PTB	O	2124.86	1716.66	1716.66	DA	REJECT - DUP DOS
14	North	8/29/03	Hosp4	6/2/03	1/1/00			F	PTB	O	12	12	12	DA	REJECT - DUP DOS
15	North	9/5/03	Hosp6	8/8/03	1/1/00			F	PTB	O	304	304	304	DA	REJECT - DUP DOS
16	North	9/5/03	Hosp6	8/8/03	1/1/00			F	PTB	O	89	89	89	DA	REJECT - DUP DOS
17	North	9/5/03	Hosp6	8/8/03	1/1/00			F	PTB	O	89	89	89	DA	REJECT - DUP DOS
18	North	9/11/03	Hosp6	2/4/03	1/1/00	OL		F	PTB	O	3728	361	361	DA	REJECT - DUP DOS

MISTAKE-PROOF DATA COLLECTION

People use Excel to create forms for all kinds of data collection: time sheets, scorecards, even mini-databases. Unfortunately, when they try to analyze the data with pivot tables, they soon discover that humans are very creative spellers. One hospital system spelled "Medicare" with various acronyms: MDCR, Medcr, Medicr, etc. This makes it difficult to do any data analysis or mining without a lot of cleanup. Excel's Data Validation function can eliminate the confusion.

Data Validation with Excel

The travel department of a major company asked its travel agents to track flights using dates, route codes (e.g., DEN-LAX-DEN), and destination city. Travel agents found creative ways to make the analysis difficult: incorrect dates, swapping routes with destinations, misspelling destinations, leaving the hyphen out of the route, and so on (Figure 7.22).

FIGURE 7.22

Bad travel data.

	A	B	C
1	Date	Route	Destination
2	2/20/2010	DEN-LAX-DEN	Los Angeles
3	Thursday	DEN_LAX-DEN	LA
4	21-Feb	Denver-LA	DEN LAX DEN

How can the travel group ensure that travel agents enter the data correctly? Data validation. Simply select the cells in the column and then specify a format and content for those cells.

In Excel 2000–2003, click on "Data Validation" (Figure 7.23). In Excel 2007–2010, click on the "Data" tab, and choose "Data Validation."

Excel will pop up a menu with various choices: integers, decimals, dates, times, text length, list, or custom (Figure 7.24). In the case of the travel group, it was time to mistake-proof the data-entry form.

FIGURE 7.23

Excel data validation.

FIGURE 7.24

"Data Validation" menu.

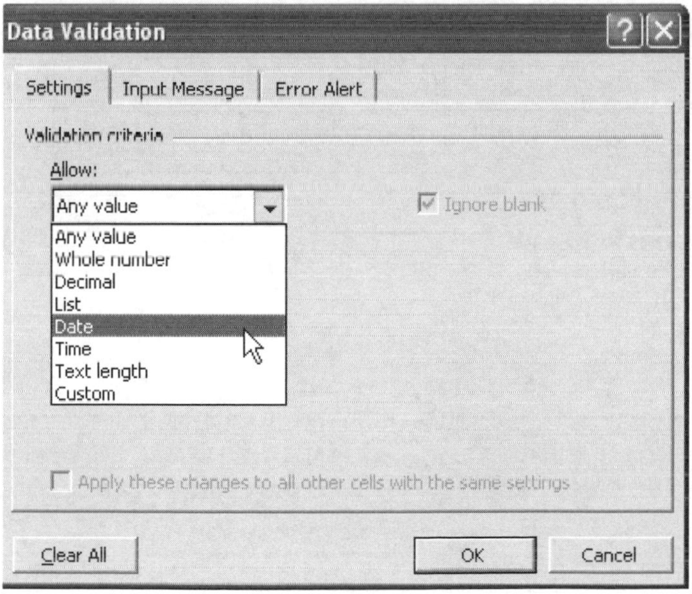

To clean up the dates, Excel's data validation can require specific formats, in this case, a date after January 1, 2010. Just select column A, "Data Validation," and specify the criteria (Figure 7.25). If a travel agent tries to put in an incorrect date, Excel will tell him or her that it's invalid (Figure 7.26).

Select column B (the route), choose "Custom," and insert a formula to check for a hyphen in character four (Figure 7.27). Excel's MID function can check the cell for a hyphen. In this case, MID looks in the cell, at character 4, for a length of one: MID(B1,4,1). Since I selected the entire column, I used the first cell, B1, as the starting point, and it will apply to all the cells. If an agent tries to put in anything that doesn't have a hyphen at character 4 (e.g., DENVER to LAX), Excel will display an error.

FIGURE 7.25

Data validation.

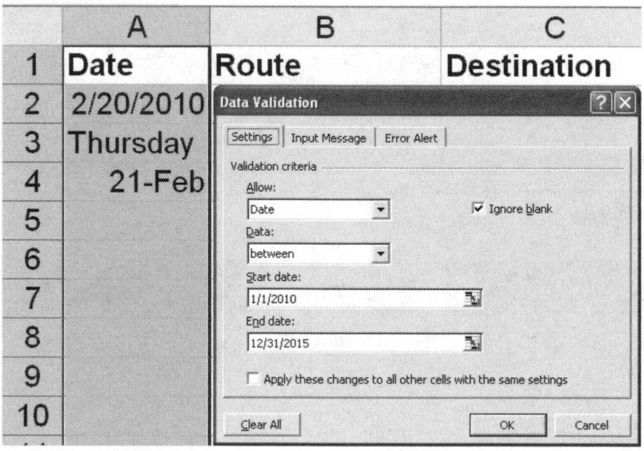

FIGURE 7.26

Invalid date message.

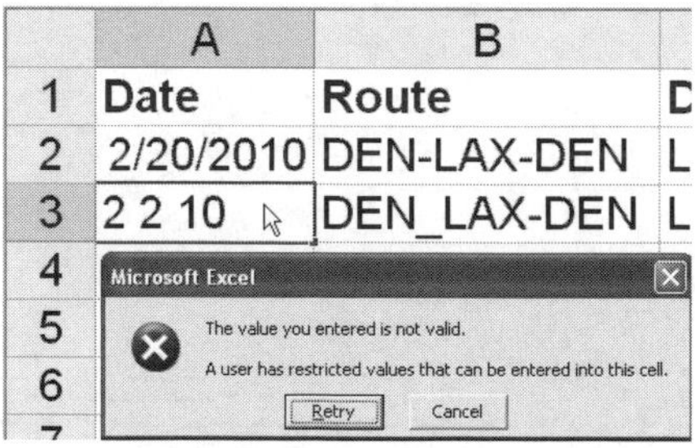

FIGURE 7.27

Using custom formats for validation.

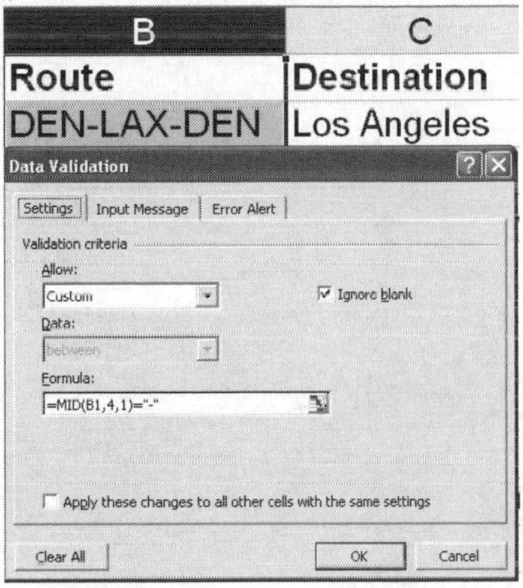

B	C
Route	**Destination**
DEN-LAX-DEN	Los Angeles

Data Validation

Settings | Input Message | Error Alert

Validation criteria

Allow:
Custom   ☑ Ignore blank

Data:
between

Formula:
=MID(B1,4,1)="-"

☐ Apply these changes to all other cells with the same settings

Clear All   OK   Cancel

The destination column is a bit more challenging. Data validation will let you specify a list of values, but travel destinations might be too varied. Excel, however, will autocomplete a cell after a few characters, so you could enter the 10 most common destinations in C2:C11 (Figure 7.28). Then, when an agent starts to type in that column, the destination will appear (e.g., San Francisco). If trips all originate from a common destination (e.g., Denver), you could enter routes as well.

Then use "Format—Row—Hide" to hide rows 2 through 11 (Figure 7.29). With rows 2 through 11 hidden, agents will be

FIGURE 7.28

Using the Excel autofill function.

	A	B	C
1	Date	Route	Destination
2		DEN-LAX-DEN	Los Angeles
3		DEN-SFO-DEN	San Francisco
4		DEN-SEA-DEN	Seattle
5		DEN-ORD-DEN	Chicago
6		DEN-ATL-DEN	Atlanta
7		DEN-HOU-DEN	Houston
8		DEN-DFW-DEN	Dallas
9		DEN-JFK-DEN	New York
10		DEN-IAD-DEN	Washington DC
11		DEN-BOS-DEN	Boston
12		DEN-SFO-DEN	San Francisco
13			San Francisco

Format hide rows.

prompted with destinations and routes when they begin to type.
Now let's look at another way to do this using a list.

Time-Sheet Example

Imagine a human resources (HR) staffer trying to get valid time
sheets using Excel (Figure 7.30). The hours are rounded to the near-
est half hour. But employees keep putting in values such as 4:30
instead of 4.5. Here's how easy it can be to solve this problem with
data validation.

List of times in half-hour increments.

	A	B	C	D	E	F	G	H	I
1	Employee	Mon	Tue	Wed	Thu	Fri	Sat		Hours
2									0
3									=I2+0.5
4									1
5									1.5
6									2
7									2.5
8									3

First, in an empty column, enter 0 in the first cell (I2), then a formula (=I3+0.5) in the next cell, and copy/paste the formula down to get 24 hours (see Figure 7.18). Then select columns B:G and "Data Validation—List" to specify the source list (=$I$2:$I$50) (Figure 7.31).

This will add a drop-down list to every cell so that employees can type a valid time (e.g., 2.5) or select a valid time (Figure 7.32). Then just hide column I and save the workbook.

FIGURE 7.31

Data validation using a list.

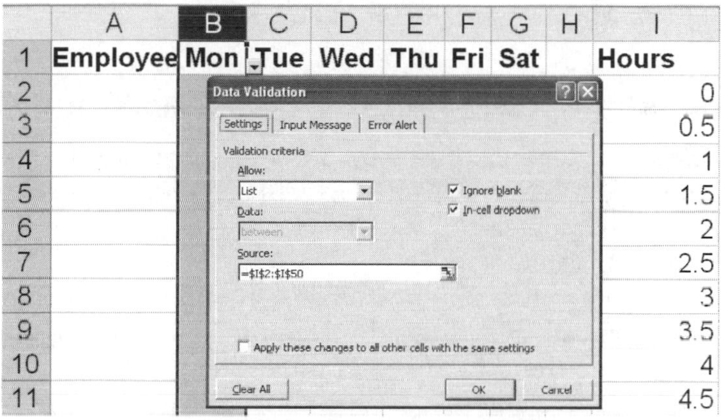

FIGURE 7.32

Drop-down list of values.

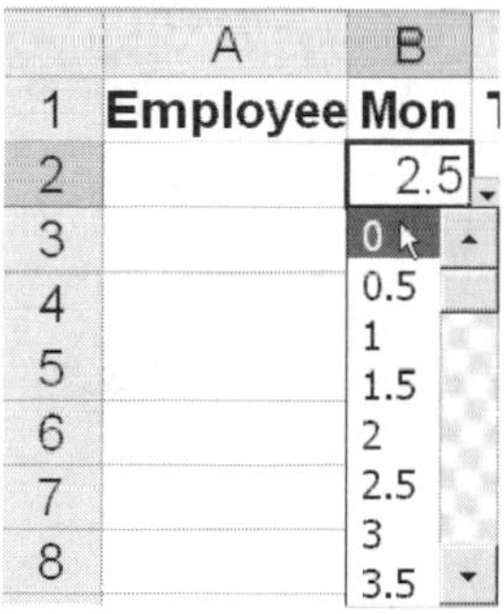

Custom Prompts

There are two other tabs on the data validation menu: "Input Message" and "Error Alert."

Input Messages.   If you want to prompt people every time they enter data into a cell, "Input Message" can help them format it correctly (Figure 7.33). This can be a little intrusive after people learn how to use the data sheet.

Error Messages.   Or you can specify an error message that appears only when users enter an invalid value (Figure 7.34). If a user tries to enter four hours and thirty minutes as 4:30, he or she will get the error message shown in Figure 7.35.

Successful Six Sigma Projects Need Good Data

Gathering consistent, error-free data is one of the keys to process improvement. To create powerful Excel-based tools and improvement stories without a lot of data cleanup, you will need data entered in a consistent way.

Excel's data validation functions will train users to enter data correctly. Users can spend days in training, or Excel can just force them to learn the right way to enter data. Using Excel is faster and more effective. Mistake-proof your data collection. It's just this easy.

FIGURE 7.33

Input message.

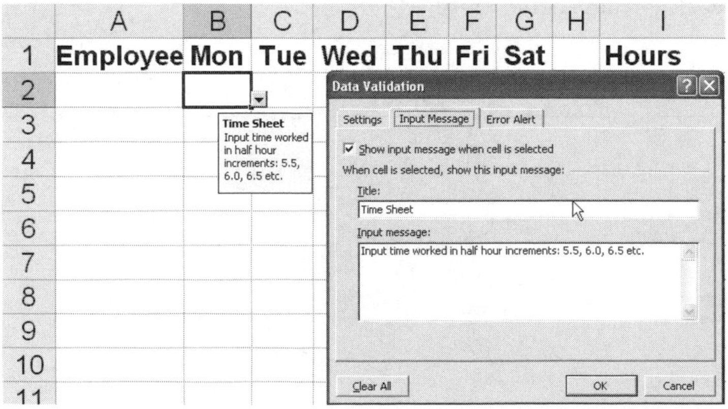

FIGURE 7.34

Data validation error message.

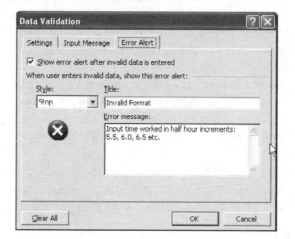

FIGURE 7.35

Invalid format message.

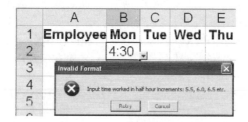

TROUBLESHOOTING PROBLEMS

Users have three types of questions when using the QI Macros:

1. *Statistical process control questions.* What chart should I use? If you use the Control Chart Wizard in the QI Macros, the software will choose your chart for you. Otherwise, most of these SPC questions are answered on our Web site at www.qimacros.com/spcfaq.html.

2. *Excel questions.* How do I enter my data? Why don't I get the right number of decimal places? and so on. Most of these are answered at www.qimacros.com/excelfaq.html.

3. *QI Macros/Excel/Windows support issues.* Most of these are answered at www.qimacros.com/techsupport.html.

Here are some common issues:

- *How do I set up my data?* See examples in QI Macros test data.
- *Decimal points (e.g., .02).* Excel stores most numbers as general format. To get greater precision, simply select your data and go to "Format—Cells—Number" to specify the number of decimals. Then run your chart.
- *Headers shown as data.* Are your headers numeric? If so, you need to put an apostrophe (') in front of each heading.
- *No data (one cell), too much data (entire columns/rows), or the wrong data selected.* Are just the essential data cells highlighted?
- *Data in text format.* Are your numbers left aligned? To convert to numbers, simply put the number 1 in a blank cell, select "Edit—Copy," then select your data, and choose "Paste—Special—Multiply."
- *Hidden rows or columns of data.* Users sometimes hide a column or row of data in Excel (e.g., columns show A, B, and then F). If you select A–F, you get all the hidden data too!
- *Data in the wrong order.* Some of these macros require two or more columns of data. The p chart expects (1) a heading, (2) the number of defects, and (3) the sample size. If columns 2 and 3 are reversed, the chart won't work properly.
- *To uninstall the Macros.* Simply delete all *.xla and *.xlam files from Excel's startup folder at c:\program files\ microsoft office\office(10, 11, 12, 14)\xlstart.

TECHNICAL SUPPORT

If you're still having problems, check out www.qimacros.com/ techsupport.html or e-mail your Excel file and problems to support@qimacros.com. Include the version number and service pack of Excel and Windows or MacOS.

E-mail qimacros@aweber.com for a *free* supplemental e-mail course on the QI Macros. Sign up for the free monthly QI Macros Webinar at www.qimacros.com.

CHARTJUNK

I recently stumbled over a book called *Visual Explanations* (Graphics Press, 1997), by Edward R. Tufte. The *New York Times* calls Tufte "the Leonardo da Vinci of data." The author says that there are right ways and wrong ways to show data; there are displays that reveal the truth and displays that do not.

Having exhibited at National Association of Healthcare Quality (NAHQ) and Institute for Healthcare Improvement (IHI) conferences, I've seen hundreds of improvement projects displayed around the exhibit hall by attendees. Few use charts; most just use words. The charts used are often incorrect for the type of data being shown.

The right charts and data can tell an improvement story quickly and easily. Words take too long. In 2010, about 60 percent of the Baldridge Award applicants were healthcare companies. One of the things a Baldridge examiner told me is that after looking at hundreds of applications, most improvement stories tell the *before* picture, but few show the *after improvement results using control and Pareto charts.*

The Right Picture Is Worth a Thousand Words

Information displays should serve the analytical purpose at hand. Here are some of Tufte's insights:

- *Numbers become evidence by being in relation to something.*
- *The disappearing legend.* When the legend on a chart is lost, the insights can be lost as well.
- *Chartjunk.* Good design brings absolute attention to data. Bad design loses the insights in the clutter.
- *Lack of clarity in depicting cause and effect.*
- *Wrong order.* A fatal flaw can be in ordering the data. A time series (i.e., a control chart) may not reveal what a bar chart (i.e., a histogram) might.

I usually draw as many different charts from the same data as I can to see which one tells the best story. You should too. Every picture tells a story, but some pictures are better than others at telling the story. The QI Macros make it easy to draw one chart after another so that you can quickly discard some of them and select others that engage the eye in the real issues.

As Tufte would say: Don't let your charts become *disinformation.* There's enough of that in the world already.

GET THE IDEA?

Chartjunk is a form of disinformation. It confuses the reader. Clean up your charts. Get rid of unnecessary clutter. Choose the right kind of chart for your data, and you'll go a long way toward motivating readers to understand and align with the business case presented.

CHAPTER 8

Is There an Improvement Project in My Data?

Lean Six Sigma professionals often talk about picking the "low-hanging fruit," but what if the low-hanging fruit is *invisible*. I've noticed that one of the biggest challenges improvement professionals face is figuring out how to develop improvement stories from the volumes of data that reside in financial accounting systems, electronic medical records, and other data sources. In most cases, the data look like Figure 8.1. They show dates, physicians, diagnoses, patient age, length of stay (LOS), adverse events, and discharge status. Do some physicians have more complications, higher costs, or longer lengths of stay?

These kinds of data remind me of the old joke about the father who finds his daughter digging through a pile of horse manure. When he asks her what she's doing, she replies: "There has to be a pony in here somewhere!"

Is there an improvement pony in your pile of data? With data such as these, you will want to do some data mining with Excel to find the hidden low-hanging fruit. Every multimillion-dollar improvement project I've worked on started from these kinds of data.

DATA MINING WITH EXCEL AND THE QI MACROS PIVOTTABLE WIZARD

Excel's PivotTable function can count or sum the number of times a word or phrase occurs with a value. It's pretty remarkable what

F I G U R E   8.1

Denied claims data.

	A	B	C	D	E	F	G	H	I	J	K
1	Region	POST DATE	ENT	ADM DATE	DIS DATE	AS	COS	FC	IN1	PT	DENIED CHARGES
7	North	11/5/02	Hosp4	8/6/01	1/1/00	OL		F	PTB	E	3224.83
8	North	11/20/02	Hosp5	4/15/02	1/1/00	OL		F	PTB	O	3291.76
9	North	11/27/02	Hosp1	5/13/02	1/1/00	OL		F	PTB	O	13845.9
10	North	11/27/02	Hosp4	9/16/02	1/1/00			F	PTB	O	1151
11	North	12/11/02	Hosp6	9/24/02	1/1/00			F	PTB	O	797
12	North	6/18/03	Hosp7	4/15/03	1/1/00			C	OT1	O	215
13	North	6/25/03	Hosp5	7/26/01	1/1/00	OL		F	PTB	O	2124.86
14	North	8/29/03	Hosp4	6/2/03	1/1/00			F	PTB	O	12
15	North	9/5/03	Hosp6	8/8/03	1/1/00			F	PTB	O	304

it can summarize. Most people don't know how to use it. Thus I created a PivotTable Wizard in the QI Macros to make pivot tables simple. Simply select up to four headings using the mouse and Alt key (e.g., Region, Entity, Admit Date, and Denied Charges, using the data in Figure 8.1). Then choose the QI Macros "Data Transformation—PivotTable Wizard" to create the pivot table (Figure 8.2).

Data Mining with Excel

Using Excel's PivotTable function, it's easy to analyze these data to find improvement stories. Here's the process:

F I G U R E   8.2

Pivot table of denied claims.

	A	B	C	D	E	F	G	H	I
1	Region	(All)							
2									
3	Sum of DENIED CHARGES	ENT							
4	ADM DATE	Hosp1	Hosp2	Hosp3	Hosp4	Hosp5	Hosp6	Hosp7	Grand Total
5	3/28/00			387.48					387.48
6	4/25/00			379.62					379.62
7	3/13/01			6908.98					6908.98
8	7/24/01		311.16						311.16
9	7/26/01					2124.86			2124.86
10	8/6/01				3224.83				3224.83
11	8/20/01		193.65	343.51					537.16
12	10/23/01			230.42					230.42
13	11/16/01			2186.16					2186.16
14	11/19/01			2627.84					2627.84
15	11/26/01			311.2					311.2

1. Click on any cell in the data. (Excel will automatically select all the rows and columns in your data.)

2. Click on Excel's "Data—PivotTable" (Excel 1997–2003) or "Insert PivotTable" (Excel 2007–2010), and then click "Finish" to reveal the pivot table fill-in-the-blank template (Figure 8.3).

   - Now simply use the mouse to drag and drop items onto the pivot table fields. Want denied charges by day? Drag "Date" into "Drop Row Fields Here," and drag "Total Charge" into "Drop Data Items Here." Then use the QI Macros to create an XmR chart of denied charges (Figure 8.4).

   - Want total charges by hospital? Drag "Entity" into "Drop Row Fields Here," and drag "Total Charge" into "Drop Data Items Here." Then sort the results in descending order by clicking on the "Total Column" and "Data-Sort-Descending." Then use the QI Macros to draw a Pareto chart (Figure 8.5).

3. If you want to see the data that make up any cell in the pivot table, just double click on the cell. If you click on the "Grand Total for Hosp2," you get the data in Figure 8.6.

FIGURE 8.3

Excel PivotTable template.

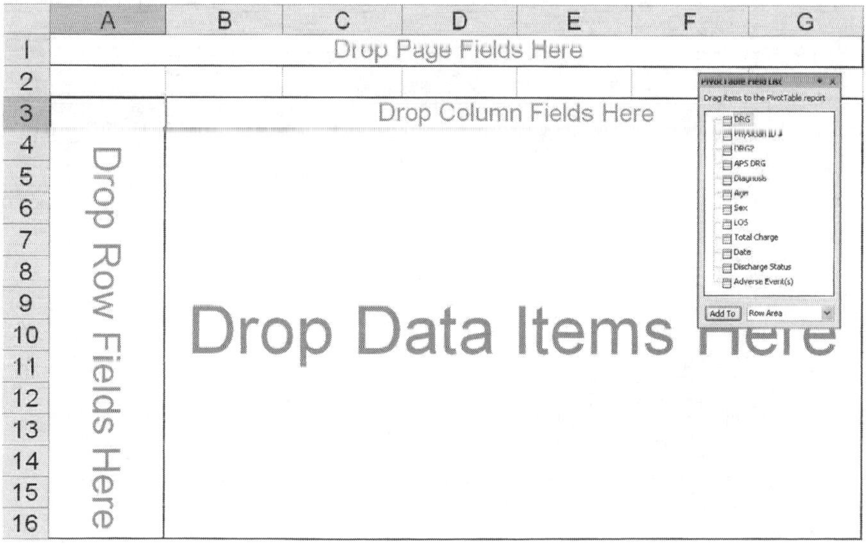

FIGURE 8.4

Total denied charges by day XmR chart.

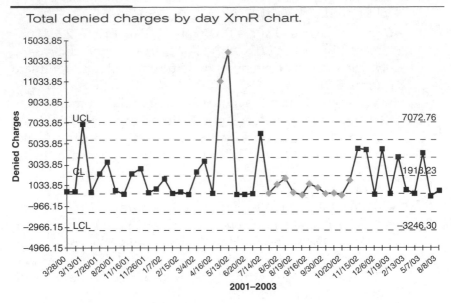

FIGURE 8.5

Denied charges by hospital Pareto chart.

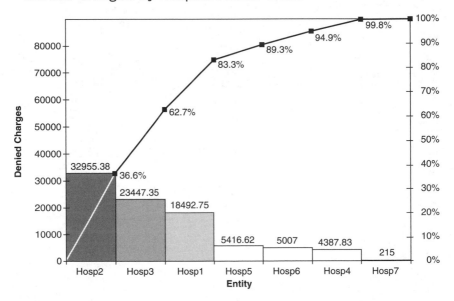

FIGURE 8.6

Denied charges for hospital 2.

	A	B	C	D	E	F	G	H	I	J	K	L	M	N	O
											DENIED				
1	Region	POST DATE	ENT	ADM DATE	DIS DATE	AS	COS	FC	IN1	PT	CHARGES	ACCT BAL	INS BAL CALC	MEMO	DESC
2	South	10/29/2002	Hosp2	7/24/2001	1/1/1900	OL		X	BAN	O	311.16	311.16	0	DC	REJECT - OVERLAPPING DOS.
3	South	11/20/2002	Hosp2	8/20/2001	1/1/1900	OL		X	BCP	R	193.65	110.4		DC	REJECT - OVERLAPPING DOS.
4	South	10/31/2002	Hosp2	12/21/2001	1/1/1900		MB	X	BAN	O	643.81	643.81	643.81	DC	REJECT - OVERLAPPING DOS.
5	South	12/10/2002	Hosp2	1/7/2002	1/1/1900	OL		3	SHD	O	1630.38	1630.38		DC	REJECT - OVERLAPPING DOS.
6	South	11/25/2002	Hosp2	2/15/2002	1/1/1900	OL	MX	X	BAN	O	372.54	372.54	372.54	DC	REJECT - OVERLAPPING DOS
7	South	6/17/2003	Hosp2	3/1/2002	1/1/1900	OL		X	UOE	O	129.5	129.5	129.5	DC	REJECT - OVERLAPPING DOS.
8	South	1/2/2003	Hosp2	3/4/2002	1/1/1900	OL		X	BAN	O	2299.71	2299.71	2299.71	DC	REJECT - OVERLAPPING DOS.
9	South	2/13/2003	Hosp2	5/6/2002	1/1/1900	OL		X	RMH	S	11045.57	279.76	279.76	DC	REJECT - OVERLAPPING DOS.
10	South	12/3/2002	Hosp2	6/20/2002	1/1/1900	OL		X	BAN	O	143.56	143.56	143.56	DC	REJECT - OVERLAPPING DOS.
11	South	12/24/2002	Hosp2	7/13/2002	1/1/1900	OL		X	BCP	E	215.4	215.4	215.4	DA	REJECT - DUP DOS (MULTIPLE
12	South	1/28/2003	Hosp2	7/29/2002	1/1/1900	OL		X	BCR	O	299.51	299.51	299.51	DC	REJECT - OVERLAPPING DOS.
13	South	1/3/2003	Hosp2	8/5/2002	1/1/1900	OL		3	SHD	O	1150.89	1150.89	1150.89	DC	REJECT - OVERLAPPING DOS.
14	South	3/5/2003	Hosp2	8/27/2002	1/1/1900	OL		X	UCC	R	161.31	81.73	81.73	DC	REJECT - OVERLAPPING DOS.
15	South	4/16/2003	Hosp2	8/30/2002	1/1/1900	OL		3	SHD	R	264.0	122.88	122.88	DC	REJECT - OVERLAPPING DOS.
16	South	2/11/2003	Hosp2	10/11/2002	1/1/1900	OL		X	BAN	O	270.98	270.98	270.98	DC	REJECT - OVERLAPPING DOS.
17	South	6/12/2003	Hosp2	11/15/2002	1/1/1900	OL		X	RMH	S	4535.66	4535.66	4535.66	DC	REJECT - OVERLAPPING DOS.
18	South	5/20/2003	Hosp2	11/25/2002	1/1/1900	OL		X	RMH	R	4431.53	3132.1	2670.69	DC	REJECT - OVERLAPPING DOS.
19	South	2/25/2003	Hosp2	12/6/2002	1/1/1900			X	CGH	O	157.92	157.92	157.92	DA	REJECT - DUP DOS (MULTIPLE
20	South	3/27/2003	Hosp2	1/14/2003	1/1/1900			X	OHP	R	4512.11	4512.11	2148.62	DC	REJECT - OVERLAPPING DOS.
21	South	6/9/2003	Hosp2	1/19/2003	1/1/1900	OL		3	SHD	E	245.89	245.89	245.89	DC	REJECT - OVERLAPPING DOS.

4. We then could use pivot tables to analyze just the descriptions, but it's obvious that rejects for overlapping date of service (DOS) account for all the denied charges.

Improvement Projects

I usually create control and Pareto charts as I'm mining the data. From my perspective, most improvement stories consist of using three key tools in the right order:

- Control charts
- Pareto charts (two or more levels of detail)
- Ishikawa or fishbone diagrams (Figure 8.7)

Hint: Narrow your focus. Eliminating denied charges for duplicate date of service will boost profits and reduce patient irritation.

If you go to www.qimacros.com/hospitalbook.html, you can download the QI Macros Lean Six Sigma Software 90-day trial. Use the PivotTable Wizard to create pivot tables. Use the Control Chart Wizard to draw control charts. Use the Pareto chart macro to draw Pareto charts. Click on "Fill-in-the-Blanks Templates" to access the fishbone diagram.

FIGURE 8.7

Ishikawa diagram of denied charges.

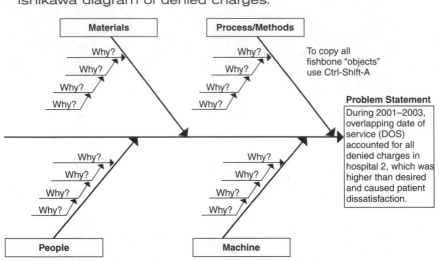

Get the Idea?

You can distill mounds of data into simple counts and sums using Excel's PivotTable function. From there, you can draw many Pareto charts and double click on the pivot table cells that show the biggest "pain" to drill down and do additional analysis.

The data is out there—Start digging!

To succeed at Six Sigma, you'll often have to analyze and summarize text data. Most companies have lots of transactions data from "flat files" such as the ones listed below, but because the data consist of words, sentences, and raw numbers, they sometimes have a hard time figuring out what to do with them.

Pivot tables can

- Count the number of times a phrase exists in a column of data (e.g., "complication")
- Count the number of times a phrase occurs in relation to another column (e.g., "physician" and "adverse events")
- Count, sum, or average numbers in relationship to another column (e.g., "average age of complications" or "total charges per physician")

Since there's so much of these data out there and so few people seem to know how to analyze them, the QI Macros include tools to simplify analyzing flat text files such as these:

- *Word Count* simply cuts each word out of every sentence in every cell you've selected and then uses pivot tables to count the occurrences and order them in descending order (maximum words: 65,536).
- *PivotTable Wizard* will take columns of data and create a pivot table that summarizes the data in the most likely manner it can devise. After years of using pivot tables, I realized that I've developed some mental rules for what types of data to put in which page, row, column, or data field for the pivot table to analyze.

(Maximum columns: one to four columns at a time. While pivot tables will handle more than four columns, I have found that four at a time is sufficient to handle most analyses. Once you've created a pivot table with four columns, you can always use the pivot table drag-and-drop functionality to add more fields.)

With Word Count and the PivotTable Wizard, you can

- Find the most frequently used words in the "MEMO" column.
- Count the number of times each doctor had a "complication" during delivery.
- Sum or average the charges per delivery by doctor.
- Count the number of deliveries for each diagnosis.

And do it easily.

WORD COUNT

Purpose: Count all the unique words in selected cells to identify patterns, trends, and Pareto patterns

A lot of interesting data are concealed in comments entered by service representatives. Word count parses the words out of sentences and paragraphs and uses pivot tables to count the occurrences of individual words and sort them in descending order. To count the words in your selection:

1. Select the cells you want to analyze (in this case, "Word Count" sheet in crosstab.xls).
2. Click on QI Macros "Data Transformation—Word Count."

Original comments look like Figure 8.8.

F I G U R E   8.8

Service rep comments in Excel.

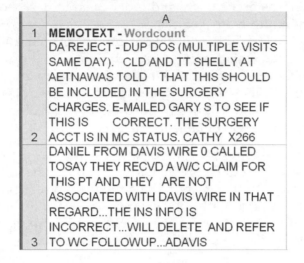

	A
1	**MEMOTEXT - Wordcount**
2	DA REJECT - DUP DOS (MULTIPLE VISITS SAME DAY). CLD AND TT SHELLY AT AETNAWAS TOLD THAT THIS SHOULD BE INCLUDED IN THE SURGERY CHARGES. E-MAILED GARY S TO SEE IF THIS IS CORRECT. THE SURGERY ACCT IS IN MC STATUS. CATHY X266
3	DANIEL FROM DAVIS WIRE 0 CALLED TOSAY THEY RECVD A W/C CLAIM FOR THIS PT AND THEY ARE NOT ASSOCIATED WITH DAVIS WIRE IN THAT REGARD...THE INS INFO IS INCORRECT...WILL DELETE AND REFER TO WC FOLLOWUP...ADAVIS

F I G U R E   8.9

QI Macros word count of key words and phrases.

	A	B	C	D	E
2	Word	Total		Two-Word Phrases	Total
3	dup	11		reject dup	8
4	dos	9		dup dos	8
5	reject	8		da reject	8
6	da	8		visits same	6
7	same	7		multiple visits	6
8	visits	6		dos multiple	6
9	multiple	6		same day	5

Word count then will parse each word out of each cell and summarize and order them using Excel's PivotTable function (Figure 8.9).

From these comments, the most likely cause of rejected claims was determined to be overlapping dates of service (DOS).

PIVOTTABLE WIZARD

> **Pivot tables are a great tool, but the user interface is too awkward for most people.**

I have found that few people know how to use Excel's PivotTable function to analyze these kinds of data. I don't know why, because it's a relatively simple drag-and-drop interface. I have to believe that it's because the user interface isn't that intuitive. That's why I created the PivotTable Wizard. Here's how you to do it step by step using my PivotTable Wizard:

Step 1: Your Data Must Have Column Headings!

As you can see in Figure 8.10, each column has a heading. The PivotTable Wizard will not run if there is a blank cell in any heading. One of the first mistakes people make is inserting blank columns to make the file more readable, and then they wonder why the PivotTable Wizard won't work.

Avoid mistakes: No blanks in column headings!

Step 2: Select the Data

You can select one to four columns of data using your mouse, or you can click on up to four column headings (e.g., Physician, Total Charge, Date, and Adverse Events), and the wizard will automatically expand the selection to include all the data in the column.

PivotTable Wizard Output. The PivotTable Wizard takes the data that you've selected and invokes Excel's PivotTable function to summarize them. Based on the content of each column, the QI Macros PivotTable Wizard figures out where to place each slice of data (page, row, column, or data field, as shown in Figure 8.11).

FIGURE 8.10

Delivery adverse event data.

	A	B	C	D	E	F	G	H	I	J	K	L
1	DRG	Physician ID #	DRG	APS DRG	Diagnosis	Age	Sex	LOS	Total Charge	Date	Discharge Status	Adverse Event(s)
2	373: Vaginal Deliver	MD10		373 3730	664.01: First-Degree Peri	26 F		2	$ 5,729	10/1/2006	Home, Self-Care	--
3	373: Vaginal Deliver	MD10		373 3730	645.11: Post Term Pregna	18 F		2	$ 9,551	10/2/2006	Home, Self-Care	--
4	373: Vaginal Deliver	MD8		373 3730	663.31: Oth Unspec Cord I	37 F		1	$ 6,976	10/2/2006	Home, Self-Care	--
5	373: Vaginal Deliver	MD8		373 3730	650: Normal Delivery	19 F		1	$ 4,589	10/3/2006	Home, Self-Care	--
6	373: Vaginal Deliver	MD1		373 3730	650: Normal Delivery	28 F		2	$11,033	10/4/2006	Home, Self-Care	--
7	373: Vaginal Deliver	MD2		373 3730	663.31: Oth Unspec Cord I	27 F		1	$ 7,002	10/4/2006	Home, Self-Care	--
8	373: Vaginal Deliver	MD3		373 3730	646.81: Oth Spec Complic	24 F		2	$ 7,190	10/4/2006	Home, Self-Care	--
9	373: Vaginal Deliver	MD3		373 3730	645.11: Post Term Pregna	21 F		2	$ 6,313	10/4/2006	Home, Self-Care	--
10	373: Vaginal Deliver	MD5		373 3730	650: Normal Delivery	19 F		1	$ 6,377	10/4/2006	Home, Self-Care	--
11	373: Vaginal Deliver	MD10		373 3730	656.61: Excessive Fetal G	22 F		1	$ 7,778	10/5/2006	Home, Self-Care	--
12	373: Vaginal Deliver	MD3		373 3730	664.01: First-Degree Peri	19 F		1	$ 6,755	10/5/2006	Home, Self-Care	--
13	373: Vaginal Deliver	MD6		373 3730	663.31: Oth Unspec Cord I	22 F		1	$ 8,369	10/5/2006	Home, Self-Care	Complication

FIGURE 8.11

Pivot table of total charges.

	A	B	C	D	E	F	G	H	I	J
1	Adverse Even	(All)								
2										
3	Sum of Total	Physician ID #								
4	Date	MD1	MD10	MD2	MD3	MD5	MD6	MD8	MD9	Grand Total
5	10/1/2006		5729							5729
6	10/2/2006		9551					6976		16527
7	10/3/2006							4589		4589
8	10/4/2006	11033		7002	13503	6377				37915
9	10/5/2006		7778		6753		15661			30192
10	10/6/2006			6464	6425	7985		7299		28173
11	10/7/2006							8344		8344
12	10/8/2006		17368				19333			36701
13	10/9/2006						20203			20203
14	10/10/2006						7328			7328
15	10/11/2006							8061	6289	14350
16	10/12/2006				33551					33551

Based on my experience in creating pivot tables, the wizard puts the adverse events in the Page field, dates in the Row field, physician in the Column field, and sum of charges in the Data field.

Step 3: Picking the Best Layout for Your Data

The PivotTable Wizard works on a drag-and-drop interface. I use

- Page fields for higher-level summaries (e.g., facility or location names).
- Row fields for the most frequent heading (often dates).
- Column fields for less frequently used headings (Excel only has 256 columns available). If you select a column with too many unique words in the cells, the pivot table will overflow.
- Data items are where you drag and drop the words or numbers you want to count or summarize.

You can change data views by clicking on the pull-down arrow next to the Page field (Figure 8.12).

Now let's say that I wanted to analyze the charges in terms of adverse events (pregnancy didn't go as planned). I'd just drag and drop "Adverse Events" into the Page fields. The PivotTable Wizard gives me a choice of viewing all charges or just the charges with the keyword "complication," "outlier," "readmission," etc. If you double click on "Sum of Total Charge" and change it to "Average," you get the average cost per delivery.

FIGURE 8.12

Pivot table Page field selection.

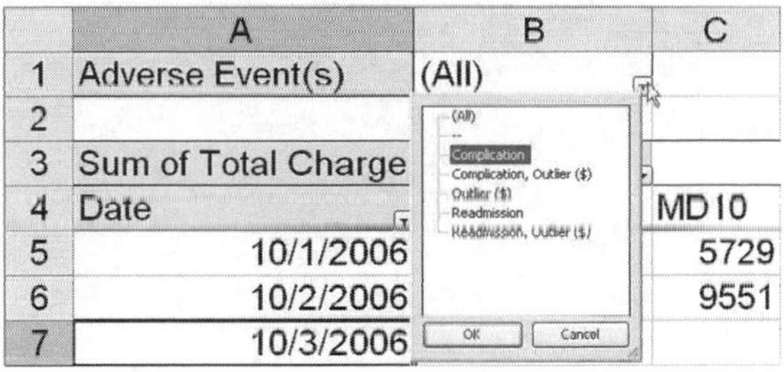

I also could change it back and group all charges into monthly charges. *How to:* Just right click on any date, and select "group" (Figure 8.13). The pivot table will group the data by month (Figure 8.14).

> **Warning! Bonehead Excel behavior: Grouping dates will not work if there is even a single blank or text cell where there should be a date.**

Now I select B4:I4 and then hold down the Control key and select "B8:I8" and run a Pareto chart (Figure 8.15).

FIGURE 8.13

PivotTable grouping menu.

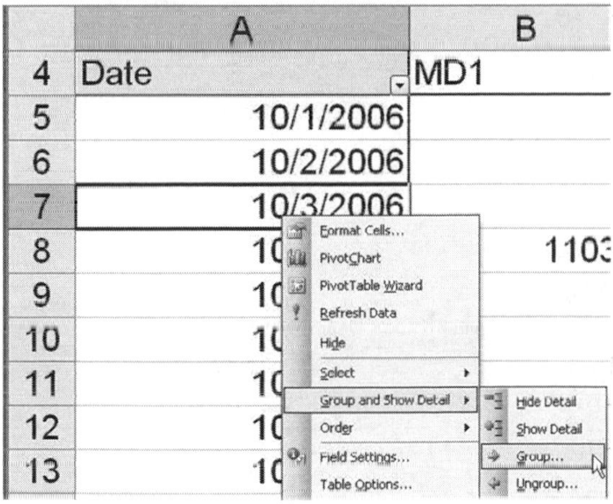

FIGURE 8.14

Grouping the pivot table by month.

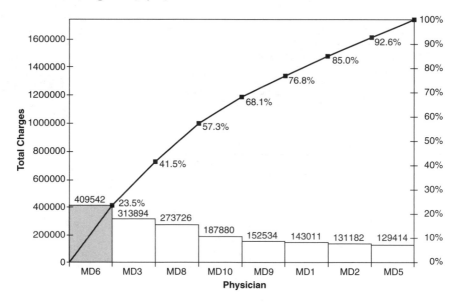

FIGURE 8.15

Total charges by physician Pareto chart.

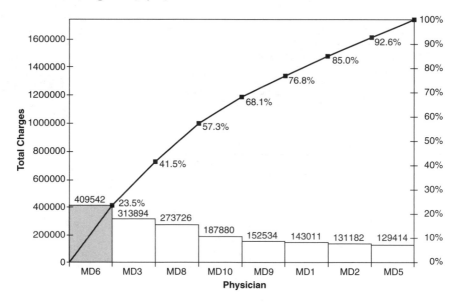

Changing the Focus

What if I wanted to change this table to count adverse events by physician?

Hint: It's easy with drag and drop.

1. Just click on date, and pull it out of the table.
2. Click on "Sum of Total Charge," and pull it out of the table.
3. Drag "Adverse Events" down into Data fields.
4. Then click on the Age field, and drag it into the "Row Fields" (Figure 8.16).

Are there more younger women with complications or are there just more younger women giving birth? Does one physician have more complications than the others? We could run a Pareto chart (Figure 8.17) to show complications by physician (MD6 has 40 percent of total adverse events, almost twice as high as his or her peers).

FIGURE 8.16

Pivot table of adverse events by physician.

	A	B	C	D	E	F	G	H	I
1	Adverse Event(s)	Complication ⌄							
2									
3	Count of Adverse Event(s)	Physician ID # ⌄							
4	Age ⌄	MD1	MD2	MD3	MD5	MD6	MD8	MD9	Grand Total
5	18			1		1			2
6	19				1	2			3
7	20			1		1			2
8	21	1				1	1		3
9	22					3			3
10	23			1					1
11	24	1		1					2
12	25					1			1
13	26		1						1
14	27	1							1
15	33							1	1
16	37			1					1
17	38		1						1
18	Grand Total	3	2	5	1	9	1	1	22

F I G U R E   8.17

Adverse events by physician Pareto chart.

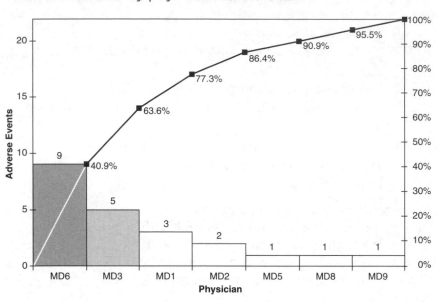

What Else?

Maybe I'd like to evaluate average length of stay (LOS) for all deliveries. Just pull the adverse events out of the table and drop in LOS and change it to an average (Figure 8.18). Not much going on here. The youngest and the oldest had a slightly longer LOS.

GET THE IDEA?

There's a wealth of information hiding in these dense flat files of words and numbers. Start using the Excel's PivotTable function to slice and dice your files (no matter how large). Then use the QI Macros to graph the results. You'll find it easy to discover the 4 percent that leads to 50 percent of the problem and start making breakthrough improvements.

As you can see from these examples, by slicing and dicing the data horizontally and vertically, you can find two or three key problem areas that could benefit from root-cause analysis.

1. Take the line-by-line error data and create a pivot table.
2. If the data are organized by date, draw a control chart of the data.

FIGURE 8.18

Pivot table of length of stay by physician.

	A	B	C	D	E	F	G	H	I	J
3	Average of LOS	Physician ID # ⌄								
4	Age	⌄MD1	MD10	MD2	MD3	MD5	MD6	MD8	MD9	Grand Total
5	14		2		2		2			2
6	15						1.5			1.5
7	17						2	1		1.5
8	18		2		1.8		1.5		1	1.5
9	19		1.33		1.3	1.3	2	1.6	2	1.6
10	20				1.4	1	1.3	1.5	2	1.4
11	21	1.5	1	1	1.7	1	1.5	1.6	1.33	1.5
12	22	1	1	1	1.7	1	1.7	2	1.5	1.6
13	23		1.33	3	2	1.7	1	1.7		1.7
14	24	1	1.5		1.5	2	2.5			1.8
15	25	1			2	1.3	1.5	2	1	1.5
16	26		1.33	1.5	1.8		1.5	1.3	1.5	1.5
17	27	2	1	1	1		1	1.4	2	1.3
18	28	2	1.25	1		2	1.3		1.5	1.4
19	29	1.5	2		2		1	1.5		1.0
20	30		1	2	1	1		1	2	1.3
21	31	2	1		2			2	1	1.6
22	32	2		1.3	2		2	2		1.8
23	33			1	1		1		1	1.0
24	34	1.5	1				2			1.6
25	35	1			1		1.5	1		1.2
26	36							1	1	1.0
27	37				2			1		1.3
28	38			1		2				1.5
29	40	2								2.0
30	Grand Total	1.53	1.36	1.4	1.7	1.4	1.6	1.5	1.43	1.5

3. Use the total columns and rows in the pivot table to draw Pareto charts with the QI Macros.

4. Then use this information to narrow your attention to one key row and column within the table. Draw the lower-level Pareto charts from these data.

5. Use the "big bars" from the lower-level Pareto charts to create problem statements that serve as the head of your fishbone diagram.

Start using the QI Macros and Excel to slice and dice your tables (no matter how large). You'll find it easy to apply the 4-50 rule and start making breakthrough improvements.

CHAPTER 9

Sustaining Improvement

Until the processes that generate the output become the focus of our efforts, the full power of these methods to improve quality, increase productivity, and reduce cost may not be fully realized.

—AIAG STATISTICAL PROCESS CONTROL
MANUAL (SECOND EDITION)

Once you've made improvements, you'll want to sustain (i.e., control) them to ensure that you stay at the new level of performance. Otherwise, you'll gradually slip back to the old levels of performance. This is why you will want a *process control system*.

A process control system of flowcharts, control charts, and/or histograms can help you to monitor and maintain your new level of performance. Process control systems consist of

1. The system—suppliers, inputs, process, outputs, and customers (QI Macros SIPOC template)
2. Charts of performance—control charts and histograms
3. Corrective actions—changes to the people, processes, machines, materials, measurement, and environment—to respond to out-of-control conditions
4. Rework—to fix defects or errors

PROCESS FLOWCHART

Once you've made an improvement, it might be a good time to develop a process flowchart or value-stream map of the process. The simplified acronym for a process is RADIO:

1. *R*epetitive—hourly, daily, weekly, monthly
2. *A*ctions—step-by-step tasks and activities
3. *D*efinable—observable and documentable (flowchart)
4. *I*nputs—measurable inputs (control charts)
5. *O*utcomes—measurable outputs (control charts)

Most processes can be diagrammed with four basic symbols:

- Start/end box
- Activity box
- Decision diamond
- Connecting arrow

Additional symbols can be added as required.

Creating a flowchart from scratch is like putting together a puzzle: It's best to get all the pieces out on the table and then try to put them in order. To do so requires flexibility, and that flexibility comes from using Post-it Notes.

> **Tip: The adhesive on Post-it Notes is better than that on other brands.**

"Swim lanes" flowcharts (Figures 9.1 and 9.2) extend the flow-charting technique to show "who does what" and the macro steps of the process.

Guidelines for constructing process flowcharts include

- Start with identifying customer needs and end with satisfying them.
- Separate the process into areas of responsibilities.
- Use Post-it Notes to lay out activities.
- Place activities under the appropriate area of responsibility.

Tips:
- Use square Post-it Notes for activities and decision diamonds.
- Draw arrows on any size Post-it Note to show the flow, top to bottom, left to right. Post-it Notes now come in arrow shapes as well.

FIGURE 9.1

Process flowchart.

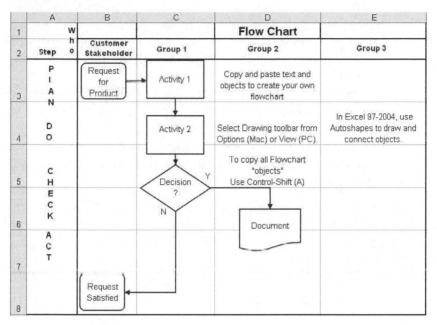

FIGURE 9.2

Swim lanes flowchart.

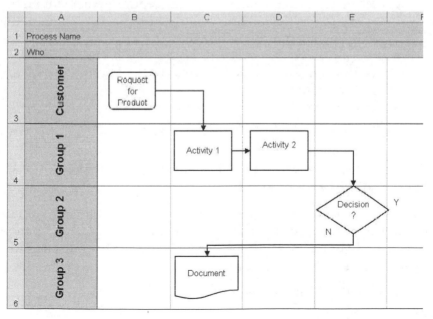

- Use smaller Post-it Notes for process and quality indicators.
- Participants often will offer activities at different levels of detail. As the higher-level process flow gets more complex, keep moving subprocesses onto microprocess diagrams.
- Critical-to-quality indicators (CTQs), which measure how well the process met the customer's requirements, go at the end of the process.
- Process indicators that predict how well the process will meet the requirements are most often placed at (1) *hand-offs* between functional groups and (2) *decision points* to measure the amount of work flowing in each direction (this is most often useful for measuring the amount of rework required).

If you go to www.qimacros.com/hospitalbook.html, you can download the QI Macros Lean Six Sigma SPC Software 90-day trial. Click on "Fill-in-the-Blanks Templates" to access the flowchart template.

Flowcharting Tar Pits

There are a few tar pits for teams to avoid:

- *Trying to show too many different kinds of process on one flowchart* (e.g., trying to show the emergency department on the same chart as the lab or trying to show procurement on the same flowchart as operations).
- *Trying to show too much detail on any one flowchart.* Use macro- and micro-level flowcharts to describe increasing levels of detail.
- *Using internal "efficiency" indicators rather than external "effectiveness" indicators based on customer requirements.*

CONTROL CHARTS FOR SUSTAINING THE IMPROVEMENT

Most hospitals will use a few main control charts—the individuals and moving range (XmR chart) for cycle times and ratios, fraction defective chart (p chart), defects chart (u chart), or g chart for "never events." Other applications include

- *Financial*—XmR charts of expenses, revenues, and so on

- *Patient, nursing, or physician satisfaction*—XmR chart of percent satisfied
- *Emergency department (ED), radiology, imaging, or outpatient wait times*—XmR chart
- *Falls per 1,000 patient-days*—XmR chart of the ratio or u chart

If you're not sure what chart to choose, just select your data and let the QI Macros Control Chart Wizard choose the chart for you.

Stability and Capability

We can use histograms to analyze process capability. Figure 9.3 shows an XmR chart of ED patient wait times to get a nursing-unit bed. The average is about 3 hours. We can use the same data to draw a histogram of ED patient wait times (Figure 9.4). To access capability, however, the process must be in statistical process control. If the process is stable and capable and meeting patient requirements, just keep monitoring. If not, it's time to crank up some improvement efforts. I doubt that patients want to be boarded in the ED for 3 hours before being moved to a nursing unit. I doubt that the hospital can afford to be on divert because of ED boarding. Solving this problem would be good for patients and profits.

	Stable—In Control	Unstable—Out of Control
Capable	Good	Analyze and correct special causes
Not Capable	Analyze and reduce common-cause variation	Correct special causes to get a stable process, then reduce common-cause variation

Reduce Variation

Once the process is stable, use process improvement to reduce the defects or deviation (adjust the process to reduce variation from the target).

Reduce the Loss

Stabilizing your process and reducing the variation will, in turn, reduce the cost of the Taguchi loss function. This will save you and

FIGURE 9.3

XmR chart of bed wait times.

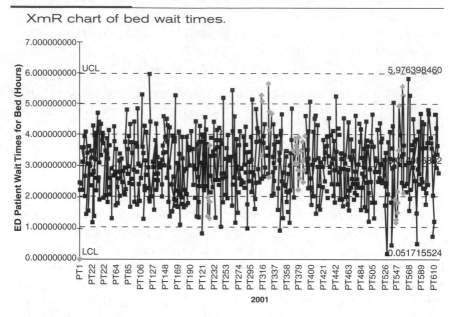

FIGURE 9.4

Histogram of bed wait times.

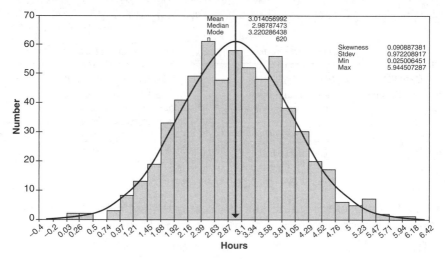

your patients time and money (rework, waste, and delay). And patients are smart. They can tell the difference between two different ED experiences, and they can tell the difference in quality between you and your competitors.

Make sure that you're in charge of who your patients return to year after year. Hitting the national benchmark isn't good enough anymore. You have to hit the patient's target value most of the time. Your patients will love you for it.

Efficiency and Effectiveness

Having a great service is essential to patient satisfaction, but you also have to deliver it in a cost-effective manner. You can only increase sales so much. There are limits to growth; it doesn't matter if you're McDonald's or Wal-Mart. To maximize profit and sustain success, you also have to trim the delays, defects, and deviation that nibble away at your profit margins.

The QI Macros started out from humble beginnings 15 years ago. Since then, I've added endless enhancements requested by customers from all over the country in everything from healthcare to the automotive industry. At a recent Institute for Healthcare Improvement (IHI) conference, a number of fans dropped by my tradeshow booth. It felt great, but I also remember that I have to endlessly improve the QI Macros and streamline their delivery. As Andy Grove of Intel fame once said, "Only the paranoid survive."

Here's my point: It's not enough to have the most innovative new product or the best customer service. If you aren't optimizing and streamlining the delivery of that product or service to reduce the excessive costs of defects, delay, waste, and rework, then your hospital will be in trouble when the bubble bursts or the fad fades.

It's easy to be seduced by easy success, but it takes clarity of focus to sustain that success. The U.S. economy is recovering, but peaks lead to troughs. Lean Six Sigma methods and tools can help you to find the lost profits in your business. Will your company be ready when the tide turns?

> With computers available, it is a waste to perform calculations by hand.
>
> —TAIICHI OHNO

Six Sigma B.C. (Before Computers)

Lately I've become concerned about how people learn statistical process control (SPC). Most trainers teach participants how to do all the calculations manually and then show them how to do them using a tool such as the QI Macros Lean Six Sigma SPC Software for Excel.

At the ASQ World Conference, I discovered a limiting belief that's been staring me in the face for almost 20 years. Everyone seems to think that you have to know how to draw a chart manually (i.e., by hand) and do all the calculations by hand to know how to do improvement or SPC. This is idiotic!

I don't think people should have to learn how to do things manually. It's like teaching a farmer how to plow a field with a plowshare when there's a brand-new tractor that can plow eight rows at a time sitting right on the edge of the field. It's like teaching a person everything there is to know about the generation and distribution of electricity before you let them turn on a light bulb. It's a waste of time.

This hallucination that you need to know the formulas to calculate a control limit on a control chart or capability on a histogram is silly.

Limiting belief: Manual = learning.

In 1991, I attended a five-day control chart class. My fellow students and I learned every formula to every chart and calculated them by hand using a calculator. In the entire five days, we spent one hour on what the chart told us. Wrong!

A friend of mine developed a video on Pareto analysis. It was 20 minutes of how to draw a Pareto chart manually and 1 minute on what the chart means. Wrong!

As I wandered around the ASQ World Conference, I noticed all the gray-haired guys like me. We learned the charts before there was software to draw the charts. We had to do it manually. We've confused correlation with causation. You don't have to draw a chart or calculate a formula manually to learn Six Sigma. We had to learn that way, but I realized that most of us had grown up B.C. (before computers).

It's like the old story that trainers often tell in Six Sigma training: Mom cuts the ham in half before she bakes it for the holidays. We ask mom, "Why?" She says, "Because grandma did it that way." We ask grandma, "Why did you cut the ham in half?" She says, "Because my oven was too small for a whole ham."

Traditions endure because we don't know any other way. Now, using the QI Macros, you can draw Pareto charts, control charts, and histograms without knowing any of the formulas. Just draw the chart and learn something. What is the chart telling you? This is the only important thing about the chart.

Isn't it time that we, the quality community, wised up and figured out that people don't have to learn the hard way? Can't we draw a chart and learn from it without knowing the math behind it?

Sure, we can layer in the formula learning later, if need be. But, right now, what is the chart telling us about how to improve? This is the only thing that matters.

People wonder why I can teach Lean Six Sigma in a day, whereas others take a week or more. Here's my secret: I don't teach my students how to do things manually. I teach them how to do things using a computer to get charts that tell them what to improve.

It used to be important to do it manually because you had to if you wanted to get results (which meant that few people ever did it). Many people feel compelled to teach this way because that's how they were taught, but it no longer adds value from my point of view. It's just a way to fill up the class time. It's a way to turn a one-day class into a five-day class.

Employees are too busy to waste time learning more than they need to know. We no longer have the luxury of learning everything there is to know before we do anything. We only have time for the essence.

The 4-50 Rule. Four percent of the knowledge about any subject will give you half the benefit. The more you teach beyond this point, the more diffused, esoteric, and seemingly complex the knowledge becomes. If you teach someone everything, they will have no idea what's important and what isn't. They know it all, but they know nothing. They have too many choices to take action effectively. From a Lean perspective, this is classic *overproduction*—teaching the *long tail* of SPC tools that people rarely use.

Let software do the hard work accurately, which will free you up to do the important work of analyzing the charts and making improvements. *Stop majoring in minor things.* Juran said it well, "The vital few versus the trivial many." This applies to knowledge as well as improvements.

Once you've learned the essence, it's easy to add to that body of knowledge. When you've learned the whole body of knowledge in one shot, it's hard to decide which portion to use and when.

Choosing a Control Chart

With the recently added I-MR-R chart, there are now 14 control charts in the QI Macros. How do you know which one to use? When I'm working with data in Excel, I follow a simple strategy for selecting the right chart based on the *format* of the data themselves. There are three formats I look for:

1. A single row/column
2. Two rows/columns with a numerator and a denominator
3. Two or more rows/columns containing multiple observations from each sample

Single Row/Column. If you have only a single row/column of data, there's only three charts you can use:

- *c chart (attribute or counted data).* It's always an integer (e.g., 1, 2, 3, or 4 injuries per month).
- *XmR chart (variable or measured data using a mean or median for the centerline).* It usually has decimal places (e.g., 4.75 falls/1,000 patient-days). See Figure 9.9 for an example of an XmR chart of divert hours.
- *XmR trend chart for variable data that increase (e.g., rising costs owing to inflation).*

So which one should you choose? If you're counting indivisible things such as defects, people, cars, or injuries, then choose the c chart. If you're measuring things such as time, length, weight, or volume, choose the XmR (individuals) chart. Look for these patterns in the data, and then select the chart.

Two Rows/Columns. If the data have a numerator and a *denominator that varies* (e.g., medication errors/orders, falls/1,000 patient-days, C-sections/deliveries, or denied claims/claims), then you will want to use the

- *p chart (one defect per patient or product).* Example: C-sections/deliveries.
- *u chart (one or more defects per patient or product).* Example: falls/1,000 patient-days.

How can you tell which one to use? I ask myself, "Can this widget have more than one defect?" For example, "Can one patient have more than one fall?" If yes, use the u chart; otherwise, use the p chart.

Two or More Rows/Columns of Variable Data. Healthcare doesn't use these charts very often. They are mainly used in manufacturing. If you have two or more rows or columns of variable data (time, weight, length, width, diameter, or volume), then you can choose one of four main charts:

- XbarR (average and range, 2 to 5 rows/columns per sample)
- XMedianR (median and range, 2 to 5 rows/columns per sample)
- XbarS (average and standard deviation, 5 to 50 rows/columns per sample)
- I-MR-R [average, moving range (between subgroups), and range (within subgroups, 2 to 50 rows/columns per sample)]

Your data should look like Figure 9.5.

You can run the XbarR, XMedianR, XbarS, or I-MR-R chart on these data. The XbarR uses the average as the measure of central tendency. The XMedianR uses the median. If you have more than five samples per period, then the XbarS probably will be the most robust chart for your needs. You also can use the XbarS if your data have a varying number of samples per period. The I-MR-R chart is like a combination of an XbarR and XMedianR; it measures the variation *within* subgroups with the range chart and variation *between* subgroups using the moving-range chart.

FIGURE 9.5

X chart data.

	A	B	C	D	E
1		Staffing Variance by Shift			
2	Days	7-3	3-7	7-11	11-7
3	1	-1.50	0.00	0.00	0.00
4	2	-0.50	-0.50	-0.50	0.00
5	3	0.00	0.00	0.00	0.00
6	4	0.00	0.00	0.00	0.00
7	5	0.00	-1.00	0.00	0.00
8	6	-0.50	-0.50	0.00	0.00
9	7	0.00	0.50	0.00	0.00

Again, look for these patterns in your data, and then select the chart.

The g Chart. Geometric median and time between control charts can track rare events such as wrong-site or wrong-patient surgeries in a hospital (Figure 9.6). Hospitals use these charts to track "never events"—things that should *never* happen, but do.

The np Chart. There's one chart that I've left to last because I rarely find situations where it applies. The np chart is like the p chart except that the sample sizes are constant. In healthcare, sample sizes are rarely constant. The data look like Figure 9.7. Again, look for these patterns in your data, and then select the chart.

If you go to www.qimacros.com/hospitalbook.html, you can download the QI Macros Lean Six Sigma SPC Software 90-day trial. Use the Control Chart Wizard to draw control charts.

Other Control Charts. There are many other forms of control charts for various applications: short-run charts, ANOM, CUSUM, EWMA, moving average, Levey-Jennings, and Hotelling.

FIGURE 9.6

g and t charts.

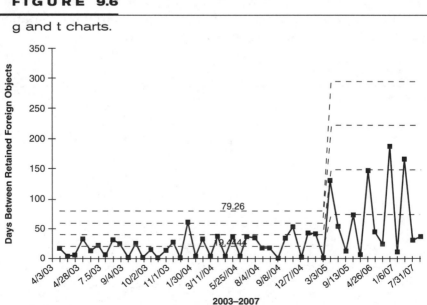

FIGURE 9.7

np chart data.

	A	B
1	Orders with Errors	
2	27	
3	19	n=60
4	18	
5	16	
6	16	
7	12	

- *Short-run charts.* What if you make only three of this product and five of that one? There's never enough data to do a full control chart. Short-run charts analyze the deviation from nominal (DNOM) target for each different product.
- *Analysis of means (ANOM).* These control charts show variation from the mean. They are used mainly used for experimental, not production, data.
- *Cumulative sum (CUSUM).* These control charts detect small process shifts by analyzing deviation from a target value.
- Exponentially weighted moving average [EWMA; a.k.a. geometric moving average (GMA)]. These charts are effective at detecting small process shifts, but they are not as effective as X charts for detecting large process shifts.
- *Moving average.* These charts can be more effective at detecting small process shifts than an XmR chart. The EWMA chart may be more effective than the moving-average chart.
- *Levey-Jennings average and standard deviation.* These charts are used extensively in laboratories.
- *Hotelling.* What do you do if you need to control two things simultaneously, such as vertical and horizontal placement of a drilled hole? Hotelling charts will assist in controlling these *multivariate* kinds of situations.

I have yet to use any of these charts in healthcare other than the g chart, but obviously, some people have advanced applications for them. I recommend getting familiar with the X, p, and u charts before turning to these other types of charts.

Summary. So, just recognizing patterns in your data can make it easier to pick the right control chart, and the QI Macros Control Chart Wizard will do this for you.

If you learn to look for these patterns in your data, it will make it easier to choose the right control chart. And it's so easy to draw these charts with the QI Macros that you can draw them and throw them away if they aren't quite right.

Stability Analysis

Once you've got a control chart, then what do you do? Processes that are out of control need to be stabilized before they can be improved using the problem-solving process. Special causes require immediate cause-effect analysis to eliminate variation.

The diagram in Figure 9.8 will help you to evaluate stability in any control chart. Unstable conditions can be any of the following:

- Any point above the upper centerline (UCL) or below the lower centerline (LCL) (Figure 9.9)
- Two of three points between 2 sigma and the control limits
- Four of five points between 1 and 2 sigma (see Figure 9.9)
- Eight points in a row above or below the centerline
- Six points in a row ascending or descending (i.e., a trend)
- And there are several other rules, called *Nelson rules*, that detect other statistically unlikely conditions.

FIGURE 9.8

Stability analysis rules.

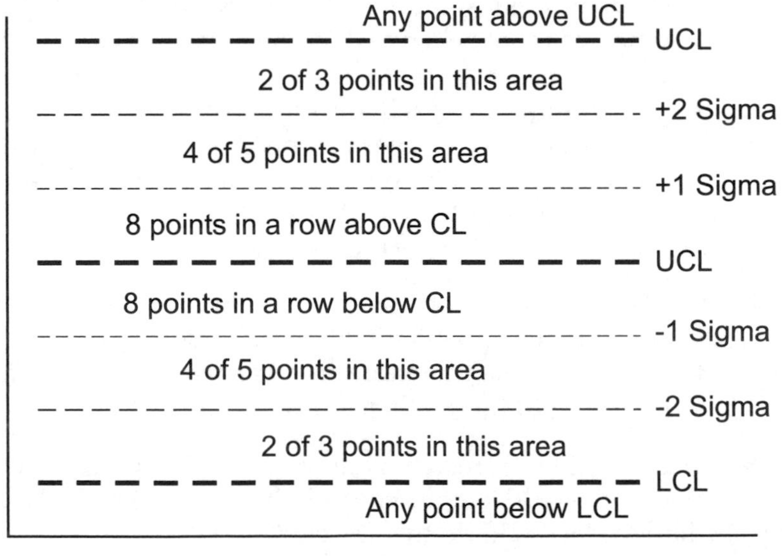

FIGURE 9.9

XmR chart of divert hours.

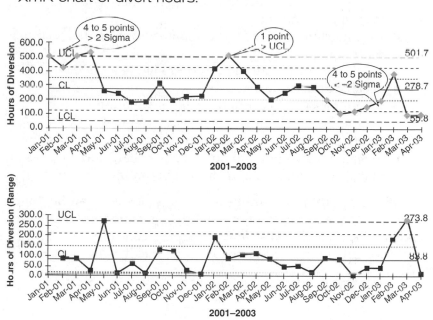

2001–2003

Any of these conditions suggests an unstable condition may exist. Investigate these special causes of variation with the fishbone diagram. Once you've eliminated the special causes, you can turn your attention to using the problem-solving process to reduce the common causes of variation. You can download my SPC quick reference card from www.qimacros.com/sustainaid.pdf.

CONTROL PLAN

For companies that need more rigor in process control, consider implementing a control plan. A *control plan* is a structured method for identifying, implementing, and monitoring process controls. A control plan describes what aspects of the process, from start to finish, will be kept in statistical process control, and it also describes the corrective actions needed to restore control. Process flowcharts and failure mode and effects analysis (FMEA) documents support development of the control plan. The QI Macros include fill-in-the-blank templates for flowcharts, FMEAs, and control plans.

The control plan for any part, assembly, or deliverable identifies

- All steps in the manufacturing or service process (e.g., inserting a central line)
- Any machines used in the manufacture or delivery (e.g., central line kit)
- Service characteristics to be controlled (e.g., infections)
- Specifications and tolerances
- Techniques for measurement and evaluation
- Sample size and frequency of measurement
- Control methods (e.g., inspection, XmR chart, and so on)
- Reaction plan—what to do when the characteristic goes out of control (e.g., adjust, recheck, quarantine)

While control plans are beyond the scope of Lean Six Sigma for hospitals, it's useful to know that there's more rigor available if needed. There's a checklist for developing control plans, FMEAs, and flowcharts in the QI Macros APQP Checklist Template.

CHAPTER 10

Laser-Focused Process Innovation

So far we've looked at ways to solve problems with delay, defects, and deviation using the methods and tools of Lean Six Sigma. After Lean Six Sigma teams have sunk their teeth into a few improvement projects, they often begin to wonder if they are working on the right issues and processes. This seems to be a natural progression: from getting early success using the improvement tools to wanting to focus the improvement efforts more precisely.

Lean Six Sigma has some excellent tools to help refine your improvement focus. Most people aren't ready to use these strategic tools until they've started to understand the basic methods and tools, however.

I've also noticed that Lean and Six Sigma started out as separate methods and tools but have been on a collision course for the last few years. I've also noticed a trend in the press toward something called *process innovation*. Just as Six Sigma eclipsed total quality management (TQM), I suspect that *process innovation* will become the new catchphrase that encompasses Lean Six Sigma. Regardless of what you name it, the improvement efforts can benefit from more rigorous focus.

FOCUSING THE IMPROVEMENT EFFORT

The focusing process was originally called *hoshin planning*. I call it *laser focus*. In this chapter you will learn how to use the key tools required to laser-focus your process innovation:

- Use the voice of the customer (VOC) to *define* customer requirements.
- Develop critical-to-quality (CTQ) measures to link the VOC to your business processes.
- Create a balanced scorecard to focus and align your organization's mission with both the long- and short-term improvement objectives.
- Select and graph indicators to measure your customer requirements and the progress of the improvement effort.

The planning process feeds directly into problem solving to increase speed, quality, and cost by reducing cycle time, defects, waste, and rework.

Voice of the Customer

If I had asked my customers what they wanted, they'd have asked for a faster horse.

—HENRY FORD

The voice of the customer helps hospitals to focus their improvement efforts in ways that will achieve breakthrough improvements in speed, quality, and cost that serve the customer. Using the voice of the customer (VOC), business (VOB), and employee (VOE), you can develop a *master improvement story* that links and aligns multiple teams and improvement efforts to achieve quantum leaps in performance improvement.

Michael George speaks of understanding the *heart of the customer*, not just the head. To understand the heart, he suggests that you will want to (1) develop strong links to both the core *and the fringes* of your market, (2) study the behavior of customers to gain insights into how they are using your hospital, and (3) include customers and their knowledge throughout the development process.

The VOC analysis gathers the customer's needs and wants as a basis for establishing objectives. Only customers can create jobs. So customer satisfaction is a central theme of Lean Six Sigma. There are *direct* customers (e.g., actual patients or payers) and *indirect* customers (e.g., shareholders and government regulatory agencies). Each customer has unique requirements that can be related to your business.

All improvements involve moving from a present way of satisfying customers to a more desired method. Before we can set the improvement processes in motion, however, we first have to define our direction of movement. Where most companies and improvement teams fail is in getting properly focused. To succeed, you will want to focus on your customers' needs and follow the data.

Once you've identified your key measurements for each of these goals, set a "Big Hairy Audacious Goal" (BHAG) for improvement. Forget 10 percent improvement. Go for 50 percent reductions in cycle time, defects, costs, system downtime, decision making, and so on. Go for 50 percent improvements in financial results and customer satisfaction. I have found that when you go for 10 percent improvements, you only get 10 percent ideas. When you go for 50 percent improvements, you get 50 percent or bigger ideas, and you often get 70 to 80 percent improvements. Breakthroughs! BHAGs also force you to narrow your focus to the 4 percent of the business that will produce the biggest return on investment.

Developing the Voice of the Customer

Developing the VOC matrix (Figure 10.1) is easy, but it forces some rigor into your thinking. This is perhaps the power of Lean Six Sigma—all the tools force people to go beyond surface-level thinking into a deeper understanding of their business.

FIGURE 10.1

Voice-of-the-customer matrix.

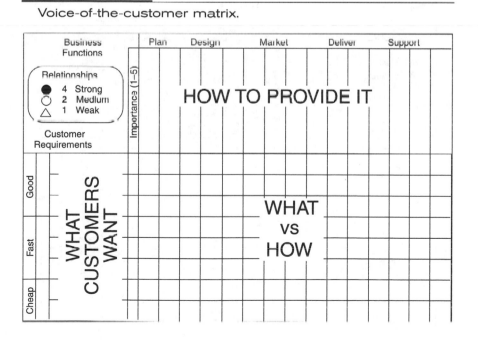

The VOC uses *the customer's* language to describe what customers want from your business. Using a restaurant as an example to elicit the participant's VOC for dining experiences, ask, "When you go into a restaurant, what do you want?"

How do restaurants provide the meals? Greet and seat, take orders, prepare and serve food, bill, and collect. What are the most important processes? Figure 10.2 explores how these requirements and processes interrelate. I have found that the VOC has some common requirements no matter what business is involved (Figure 10.3).

Customers say such things as

- Treat me like you want my business.
- Deliver products that meet my needs.
- Deliver products or services that work right.
- Be accurate, right the first time.
- Fix it right the first time.
- I want it when I want it.
- Make commitments that meet my needs.
- Meet your commitments.
- I want fast, easy access to help.

FIGURE 10.2

Restaurant voice of the customer.

	Importance (1–5)	Greet	Seat	Take Drink Order	Take Food Order	Order Supplies	Prepare Order	Serve Order	Customer Check	Take Payment
Good — Get my order right	5			●	●		●	○		
I want good food	5						●	○		
I want an accurate bill	4			●	●				●	
Give me payment options	3									●
Fast — Greet me and seat me promptly	4	●	●							
Serve me promptly	5			●	●					
Serve my food when I want it	5						●	○		
Have my check ready	4								●	
Cheap — Give me good value for money spent	4					○	●	○		
Don't waste food	3						●			
		4	4	12	12	6	16	8	8	4

Relationships: ● 4 Strong, ○ 2 Medium, △ 1 Weak. Processes: Greet, Order, Prepare, Serve, Bill & Collect. Customer Requirements = Voice of the Customer (direct & indirect).

FIGURE 10.3

Macros Voice-of-the-Customer Template.

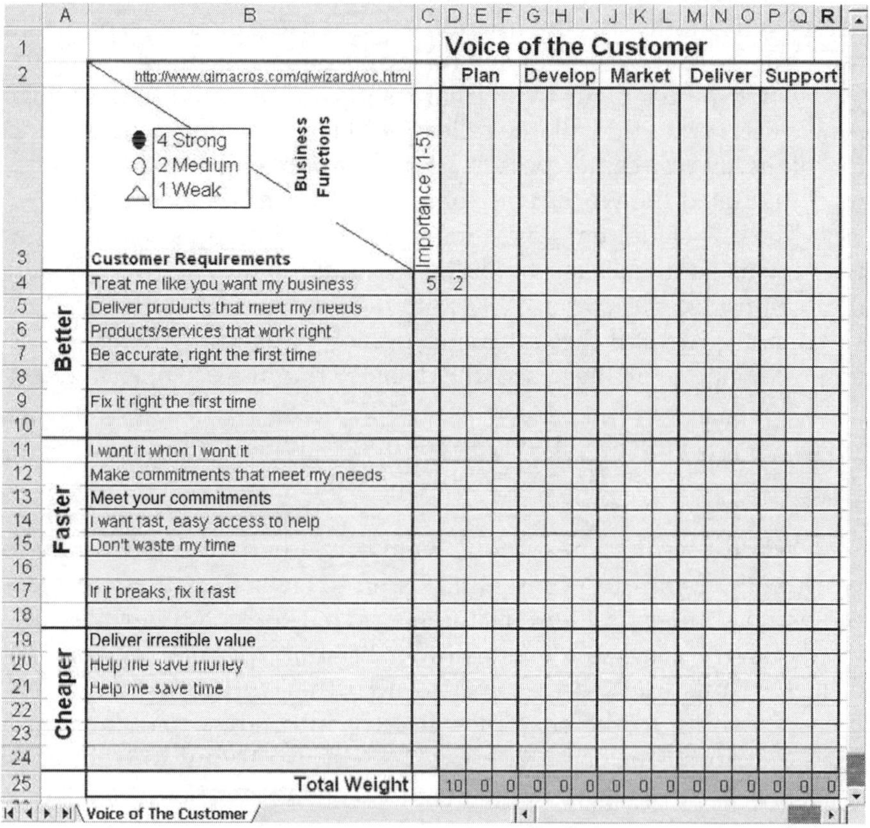

These are the most common themes I hear from customers. What are your customers saying?

- Don't waste my time.
- If it breaks, fix it fast.
- Deliver irresistible value.
- Help me save money.
- Help me save time.

These are the most common themes I hear from customers. What are your customers saying?

Speak Your Customer's Language

In Denver recently, tragedy struck a family when a 40-ton construction girder fell from an overpass onto their SUV, killing everyone. I was saddened by the tragedy, but rather than focus on the installation of the girder, my attention focused on the phone call that occurred earlier in the day that could have saved their lives. A driver with highway construction experience called to report that the girder was loose and buckling, apparently unsafe.

After the accident, TV journalists played the call for all to hear. As we listened to the call, the caller kept clearly saying "girder," and the highway call-center person kept paraphrasing the man's statement but used the word *sign*, not girder: "There's a loose sign?"

The call-center employee reported the problem as a loose sign, which was soon checked by highway maintenance staff. They didn't even notice the girder. Why not? Because they were focused on signs, not girders. If the call-center person had simply listened and entered what he or she heard, the disaster might have been averted.

Hint: Stop trying to train your customers to speak your language.

When I worked in a phone company, I had an opportunity to work in the repair call center and listen to calls. Repair-center people had a similar problem listening to what the customer was saying. They tried to teach the customer "phone speak" about central offices, trunk lines, drop boxes, and other in-house terms that meant nothing to the customer. It only infuriated customers and took up more time than it should have. Stop trying to train your customer to speak your language. It breaks rapport.

Hint: Be a parrot not a paraphraser.

In grade school, we were all taught to paraphrase what people say, but a *sign* is not a *girder*. If you truly want to listen to the VOC, in this case, a concerned citizen, you have to actually listen to and record what they say, not what you want to hear and not what you think you heard. *Parrot what they say, never paraphrase.* It builds rapport.

Hint: Don't make stuff up.

Just because you have one picture in your head doesn't mean that the person on the other end of the phone has the same picture. A picture of a loose girder in one mind doesn't equal a picture of a loose sign in another. If the call-center employee simply had written down *exactly* what the caller said, "girder," the disaster might have been avoided completely.

If you aren't sure what the customer means, don't invent a meaning, just ask: "What do you mean by 'girder'?" Then the caller might have said, "A gigantic steel beam that spans the highway," which would have changed the picture in the call-center employee's head.

Speak Your Customer's Language

If you want to communicate effectively, you have to use the words the customer gives you. Never make the customer translate what you're saying into his or her language; translate what you're saying into the customer's language so that nothing is lost in translation.

Over a decade ago, I became a master practitioner of neuro-linguistic programming (NLP). In NLP, we learned how to develop rapport by matching other people's language.

One of the principles of NLP is, "The meaning of your communication is the response you get." If your customer responds in a way that matches what you think you said, it was a good communication. If the customer responds differently, then your communication was unclear.

The language skills I learned have served me well in everything I've done. They make me a better husband because I listen to what my wife says. They make me a better consultant and supplier because I listen to what customers want and then try to deliver it in ways that match their words. I don't always get it right, but I keep working on it.

Just because we speak English does not mean that we speak the same language or that we have the same pictures, sounds, or feelings tied to any given word. We have different core values that affect our speech and five very different motivation styles that have an impact on every aspect of our communication.

Train your customer-service people to listen and connect with customers on their terms, not yours. It will make your business grow and help you to retain customers. One consultant I know worked with a major airline's written complaint department. He taught half the workers to reply to the customer in language that matched the words in customer letters. Customers who received matching-language letters increased their travel on the airline; customers receiving the normal letters did not.

I wrote a book on how to motivate everyone and a quick reference card that you can download from www.motivateeveryone.com/pdf/mejobaid.pdf. To find out more about your own motivation and communication style, you also can take my free online personality profile at www.motivateeveryone.com/nlpstyle.html. Next, you will want to figure out how to measure your ability to deliver what your customers want.

CRITICAL-TO-QUALITY INDICATORS (CTQs)

Goodness is uneventful. It does not flash, it glows.
—DAVID GRAYSON

CTQs define specific ways to *measure* the customer's requirements and to predict your ability to deliver on those requirements. All business problems invariably stem from failing to meet or exceed a customer's requirement. To begin to define the problem, you need to identify your customer's CTQ needs and a way to measure them *over time*—by hour, day, week, or month.

CTQs measure how well the product or service meets the customer's requirements. Process indicators, strategically positioned at critical handoff points in the process (Figure 10.4), provide an early warning system. For each CTQ, there should be one or more process indicators that can predict whether you will deliver what your customers require.

There are usually only a few key customer requirements for any product or service. What do your customers want? How can you measure it over time?

SIPOC

Another tool to define your current process is the SIPOC diagram (Figure 10.5). It shows your *s*uppliers, *i*nputs, main *p*rocess steps, *o*utputs, and *c*ustomers (direct and indirect). Identify your main supplier, customer, the product or service used, and the process that creates it. Begin identifying your requirements of the supplier. Then identify your customers' requirements for the product or service. What do they want in terms of good, fast, and cheap? Then, based on your customers' needs, identify how you can measure them with defects, time, or cost. Finally, identify how often you will measure by minute, hour, day, week, or month.

FIGURE 10.4

CTQ indicators and handoffs.

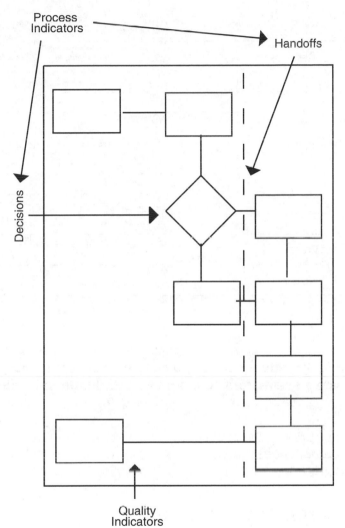

Here are examples from three different environments to demonstrate how to identify the indicators based on requirements. For an emergency department (ED), operating room (OR), nursing unit, pharmacy, and maintenance or other department, who are your main customers, products, services, processes, and customer requirements?

FIGURE 10.5

SIPOC diagram.

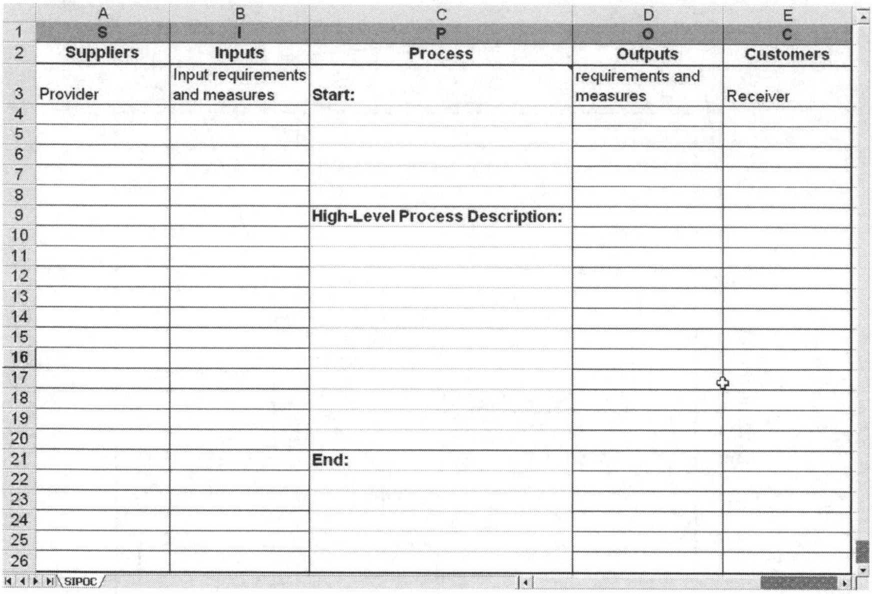

	A	B	C	D	E
1	**S**	**I**	**P**	**O**	**C**
2	**Suppliers**	**Inputs**	**Process**	**Outputs**	**Customers**
3	Provider	Input requirements and measures	Start:	requirements and measures	Receiver
4					
5					
6					
7					
8					
9			**High-Level Process Description:**		
10					
11					
12					
13					
14					
15					
16					
17					
18					
19					
20					
21			**End:**		
22					
23					
24					
25					
26					

SIPOC

> **Hint: It's often easier, as a customer, to first identify
> what you want from your suppliers and then to identify
> what your customers want from you.**

For improvement efforts to be successful, they must focus on the customers' requirements and ways to measure them in defects, time, or cost (Figure 10.6).

BALANCED SCORECARD

A *balanced scorecard* links all your efforts to ensure breakthrough improvements, not just incremental ones. The easiest way to depict this is with a tree diagram.

A balanced scorecard begins with a vision of the ideal world—fast, affordable, flawless healthcare. This vision then is linked to long-term customer requirements, short-term objectives, measures, and targets. This is a great place to involve your leadership team.

FIGURE 10.6

QI Macros measurements matrix.

	A	B	C	D
1		**Measurements**		
2		Customer/Stakeholder:	Product/Service:	
3		Customer Requirements	Measurement	Period
4		Treat me like you want my business	Customer Complaints	
5	**Better**	Deliver products that meet my needs		
6		Products/services that work right	% defective	
7		Be accurate, right the first time		
8				
9		Correct/fix it right the first time	Number/% repeat repairs	
10				
11	**Faster**	I want it when I want it	Cycle time	per product
12		Make commitments that meet my needs	% first choice commitment	
13		Meet your commitments	% commitments missed	Daily
14		I want fast, easy access to help	% calls answered < 60 sec and %calls referred	Daily
15		Don't waste my time		
16				
17		If it breaks, fix it fast	Repair cycle time	per repair
18		If it doesn't work, resolve it fast		
19	**Cheaper / Value**	Deliver irresistible value		
20		Help me save money		
21		Help me save time		
22				
23				
24				
25				
26				

MEASURES

What's Important about a Balanced Scorecard?

1. If leadership does it, the leaders will commit to achieving it.
2. It links customer needs to the improvement efforts. This clear linkage, which is often missing, helps employees and leaders to focus on the customer and align all their actions to achieve customer outcomes, not internal ones.
3. Measurements based on customer requirements provide an ideal way to evaluate performance.
4. Detailed balanced scorecards then can be developed and linked to this one by individual managers.
5. Results can be measured and monitored easily.

Long-term customer requirements invariably fall into one of three categories (from the VOC matrix):

- *Better quality*—reliability and dependability

- *Faster service*—speed and on-time delivery
- *Higher perceived value*—lower cost

Short-term objectives translate these customer "fluffy" objectives into more concrete ones that can be measured and improved to meet the targets (from indicators):

- *Better quality*—fewer defects
- *Faster service*—reduced cycle time
- *Higher perceived value and lower costs*

Targets are the BHAGs that challenge our creativity and ability. Fifty percent reductions in cycle time, defects, and costs are both challenging and achievable in a one-year period. To do so, however, requires highly focused, not random, improvement work.

QUALITY MANAGEMENT SYSTEMS

Ultimately, methods and tools such as a balanced scorecard, voice of the customer, and measurements (e.g., control charts, histograms, Pareto charts, etc.) weave together into a system for producing a quality product. Quality management systems come in many flavors: ISO9000, Capability Maturity Model Integration (CMMI), Project Management Institute (PMI), and a host of others. They all boil down to a simple series of questions:

1. Do you have a predictable process for delivering your product or service?
2. Do you follow it?
3. Are you improving it?

Look at any mature company that consistently produces a quality product, and it must have some sort of quality management system in place. While it would be nice to simply install such a system, trying to do it all at once might just kill your hospital. Consider using a crawl-walk-run approach instead. Start with a few key processes, and bring them under control. Add other processes as you gain experience. Continue until you've migrated most, if not all, of the business processes to a quality management system. It won't break the bank. It won't put too big of a strain on the business or employees. And eventually, it will deliver the desired result.

Regardless of what we call future versions of Lean Six Sigma—quality management or process innovation—business success will continue to depend on the ability to balance innovation and improvement. First, innovate to create new products; then improve to simplify, streamline, and optimize delivery of the new products. The methods and tools of Lean Six Sigma must become a part of your business mind-set if you are to compete in the global economy. It's that simple.

CHAPTER 11

Statistical Tools
for Lean Six Sigma

Some tools of Lean Six Sigma aren't graphical; they're simply analytical. Sometimes you want to be able to compare two processes or products and learn something about their quality using statistics alone. Healthcare research often uses these tools to compare the effects of two or more medications or protocols. Six Sigma practitioners can use these tools to compare performance before and after an improvement to verify its effectiveness. This falls under the category of something known as *hypothesis testing*.

When doing research or clinical trials, these statistical tools help to evaluate the efficacy of various medications or protocols. When using Lean Six Sigma to solve problems, they can help to compare the before and after performance to determine if the results are statistically significant.

HYPOTHESIS TESTING

I've come to suspect that hypothesis testing is where statistics got the nickname "sadistics." I found it confusing because it seems to use negative logic to describe everything. But it's really not that hard once you understand how it works.

Let's say that you have two medications or protocols, and you want to prove that they are (1) the *same* (i.e., equal) or (2) *different* (i.e., not equal) at a certain level of confidence. You might want to compare a control group with a test group to determine if a medication is effective. Because Lean Six Sigma is obsessed with variation and central tendencies, you might want to prove that the averages or variation is the same or different. Hypothesis testing helps you to evaluate these two *hypotheses*.

The English geneticist who dreamed this up decided that the same or equal result would be called the *null hypothesis*. Different results are known as the *alternate hypothesis*. Then, based on the analysis, you want to *accept* the null hypothesis (i.e., that the two medications or protocols are the same) or *reject* the null hypothesis (i.e., that the two medications or protocols are different). There are several tools that can help you to do this depending on whether you are most interested in the average or the variation.

HYPOTHESIS TESTING FOR VARIATION

Since variation can affect results, it's useful to determine if variation in two or more samples is the same or different. To evaluate variation statistically, use the F test or Levene's test.

F Test for Variation

If you have a single factor measured at two levels (e.g., calcium levels) and you want to know if they have the same or different variability, use the F test. An F test using two samples compares two independent sets of test data. It helps to determine if the variances are the same or different from each other. Consider the following example.

F-Test Two-Sample Example.   If you're using two different anticoagulants, you might want to know if the variances for calcium are the same or different.

- *Null hypothesis* (H_0): Variances are the same.
- *Alternate hypothesis* (H_a): Variances are different.

Now conduct a test and enter the data into Excel (Figure 11.1). Use Excel's Data Analysis ToolPak under the Tools menu or the QI Macros to conduct the F test of EDTA versus PPT. The QI Macros will prompt for a significance level (default = 0.05). The F test will calculate the results (Figure 11.2).

If you go to www.qimacros.com/hospitalbook.html, you can download the QI Macros Lean Six Sigma SPC Software 90-day trial. Use the ANOVA tools to access the F test.

FIGURE   11.1

F-test data.

1	Sample	EDTA	PPT	COL	AAS	Calcium in solution (PPM)
2	1	2.98	3.7	3.5	4.4	
3	2	2.95	3.9	4.4	5	
4	3	2.15	3.8	3.3	3.5	
5	4	3.41	4.1	2.3	3.7	
6	5	3.97	3.8	3.3	4.7	
7	6	3.86	4.4	3.9	4.1	

FIGURE   11.2

F-test results.

	A	B	C	D	E	F	G
1	EDTA	PPT	F-Test Two-Sampl	α	0.05		
2	2.98	3.69					
3	2.95	3.9		EDTA	PPT		
4	2.15	3.83	Mean	3.22	3.94		
5	3.41	4.08	Variance	0.45672	0.06436		
6	3.97	3.76	Observations	6	6		
7	3.86	4.38	df	5	5		
8			F	7.10			
9			P(F<=f) one-tail	0.025	0.051	Two-tail	
10			F Critical one-tail	5.05	7.15	Two-tail	
11			One-tail	Reject Null Hypothesis because p < 0.05			
12			Two-tail	Accept Null Hypothesis because p > 0.05			

Interpreting the F-test results

Hypothesis Test	Compare	Result
Classical method	Test statistic > critical value (i.e., $F > F_{crit}$)	Reject the null hypothesis
	Test statistic < critical value (i.e., $F < F_{crit}$)	Accept the null hypothesis
p-Value method	$p < \alpha$ Reject the null hypothesis	$p > \alpha$ Accept the null hypothesis

Since $F < F_{crit}$ (7.10 < 7.15) and $p > \alpha$ (0.051 > 0.05), we can accept the null hypothesis that the variances are equal (two-tailed) but reject the null hypothesis for one tail.

Levene's Test for Variation in Nonnormal Data

Levene's test compares two or more independent sets of test data. It helps to determine if the variances are the same or different from each other. Levene's test is like the F test. However, Levene's test is robust enough for nonnormal data and handles more than two columns of data. Consider the following example.

Levene's Test: Two-Sample Example. Using the F-test example data (see Figure 11.1), conduct a Levene's test using the QI Macros (Levene's test is not part of Excel's Data Analysis ToolPak). The Levene's test macro will calculate the results (Figure 11.3).

Interpreting Levene's test results

	Hypothesis Test	Compare Result
p-Value method	$p < \alpha$	Reject the null hypothesis
	$p > \alpha$	Accept the null hypothesis

Since Levene's $p > \alpha$ ($0.058 > 0.05$), we can accept the null hypothesis that the variances are equal.

While an F test works well on two samples of normal data, it isn't robust enough to handle nonnormal data or more than two

FIGURE 11.3

Levene's test results.

	A	B	C	D	E	F	G	H	
1	EDTA	PPT		*EDTA*	*PPT*				
2	2.98	3.69	Median	3.195	3.865				
3	2.95	3.9	Mean	3.22	3.94				
4	2.15	3.83	Variance	0.4567	0.064				
5	3.41	4.08	n	6	6				
6	3.97	3.76	df	5	5				
7	3.86	4.38		Levene's					
8			Test	4.597					
9			p	0.058	Accept Null Hypothesis because p > 0.05				
10			F-Test	a	0.05				
11			F	7.10					
12			p 1&2 tai	0.025	0.051				
13			F Critical	5.05					

samples. (Notice that Levene's p value differs from the F test's two-tailed value of 0.305; however, both cause acceptance of null hypotheses.)

Levene's Test: Four-Sample Example. Now consider all four anticoagulants (Figure 11.4). Again the $p = 0.281$, which is greater than 0.05, so we accept the null hypothesis that variances are equal from batch to batch.

HYPOTHESIS TESTING FOR MEANS

Since the mean (i.e., average) also affects results, it can be useful to evaluate whether the means of one or more samples are the same or different. There are a number of ways to do this depending on sample size: t tests, the Tukey test, and ANOVA.

t Test for Means

t tests evaluate whether the means of one or two samples are the same or different. There are several types of t tests:

- t test: single sample
- t test: two sample assuming *equal* variances
- t test: two sample assuming *unequal* variances
- t test: paired two sample for means

FIGURE 11.4

Anticoagulant data.

	A	B	C	D	E	F	G	H	I
1	EDTA	PPT	COL	AAS		EDTA	PPT	COL	AAS
2	2.98	3.69	3.54	4.41	Median	3.195	3.865	3.44	4.26
3	2.95	3.9	4.4	4.96	Mean	3.22	3.94	3.46	4.22
4	2.15	3.83	3.28	3.5	Variance	0.457	0.064	0.508	0.328
5	3.41	4.08	2.28	3.66	n	6	6	6	6
6	3.97	3.76	3.34	4.68	df	5	5	5	5
7	3.86	4.38	3.92	4.11		Levene's			
8					Test	1.367			
9					p	0.281			
10					a	0.05			
11					Accept Null Hypothesis because p > 0.05				

t Test: Single Sample

A *t* test using one sample compares test data to a specific value. It helps to determine whether the sample is greater than, less than, or equal to the value. Consider the following example (in QI Macros test data/anova.xls).

*Central Line Bloodstream Infections **t** Test Example.* Let's say that you want to know that the improvement to reduce bloodstream infections is statistically below the baseline. So we develop a null hypothesis (H_0) that central line blood stream infections (CLBSI) is less than the baseline of 4.85 and the alternate hypothesis (H_a) that CLBSI is greater than or equal to 4.85:

- $H_0 < 4.85$ infections per 1,000 line-days
- $H_a \geq 4.85$ infections per 1,000 line-days

Now take the CLBSI rate after an improvement and enter the data into Excel (column A in Figure 11.5). Then select the data with the mouse, and click on the QI Macros menu to select the one-sample *t* test (one-sample *t* tests are not available in Excel's Data Analysis ToolPak). The QI Macros will prompt for a confidence level (default = 0.95, which is the same as a significance of 0.05) and test the mean/average (in this case 4.85). The *t*-test one-sample macro will calculate the results (columns B and C in Figure 11.5).

FIGURE 11.5

Single-sample *t*-test data.

	A	B	C	D	E	F	G
1	Central Line BSI Rate per 1000 line days	t-Test 1-sample					
2	3.26	Test Mean	4.85				
3	2.59	Confidence Leve	0.95				
4	2.26	N	32				
5	2.61	Average	2.16				
6	2.52	Stdev	0.468				
7	2.77	SE Mean	0.083				
8	2.62	T	32.466				
9	2.00	TINV	1.696				
10	1.68	p - One sided	0.000	Reject Null Hypothesis because p < 0.05			
11	1.98	p - two sided	0.000	Reject Null Hypothesis because p < 0.05			

Interpreting the t-test one-sample results

If	Then
Test statistic > critical value (i.e., $t > t_{crit}$)	Reject the null hypothesis
Test statistic < critical value (i.e., $t < t_{crit}$)	Accept the null hypothesis
$p < \alpha$	Reject the null hypothesis
$p > \alpha$	Accept the null hypothesis

Since the null hypothesis is that CLBSI is less than 4.85, this is a one-sided test. Therefore, use the one-tail values for your analysis.

Note: The two-sided values would apply if our null hypothesis were that H_0: mean = 4.85 hours.

Since the t statistic > t_{crit} (32.466 > 1.696) and $p < \alpha$ (0.000 < 0.05), we can reject the null hypothesis that CLBSI is greater than or equal to 4.85. We can say that we are 95 percent confident that CLBSI is less than 4.85.

Customer Service t-Test One-Sample Example. Let's say that you want to know if wait times in an emergency department (ED) are not greater than 3 minutes at a 95 percent confidence level:

- $H_0 \leq 3$ minutes
- $H_a > 3$ minutes

Observers collect wait times. This gives us the data we need to test the hypothesis (Figure 11.6).

FIGURE 11.6

Single-sample t-test results.

	A	B	C
1	Customer	t-Test 1-sample	
2	4	Test Mean	3
3	0	Confidence Level	0.95
4	0	N	43
5	0	Average	4.465116
6	10	Stdev	5.346778
7	1	SE Mean	0.815376
8	3	T	1.79686
9	10	TINV	1.681951
10	4	p - One sided	0.039776
11	2	p - two sided	0.079552

The one-sided $p < \alpha$ [0.039776 is less than 0.05 (1–0.95)], so we must reject the null hypothesis that wait times are less than or equal to 3 minutes. We can say that we are 95 percent confident that wait times are greater than 3 minutes.

t Test: Two Sample Assuming *Equal* Variances

Using the *F*-test data from Figure 11.1, you might want to know if the calcium levels are the same or different between EDTA and PPT.

Define the Null and Alternate Hypotheses

- The null hypothesis H_0 is that the mean difference $(x_1 - x_2)$ = 0, or in other words, the means are the same.
- The alternative hypothesis H_a is that the mean difference is less than or greater than 0, or in other words, the means are not the same.

Conduct an F *Test to Determine if Variances Are Equal.*   Since the two recipes aren't "paired" or dependent in any way, the first step is to determine whether the variances are equal to determine which *t* test to use. Since we've already done this, we know that the variances are considered equal for a two-tailed distribution.

Run the t *Test Assuming Equal Variances.*   Now select the data with the mouse, and click on the QI Macros menu to select the two-sample *t* test (or for Excel's Data Analysis ToolPak, *t* test two samples assuming equal variances). The QI Macros will prompt for a significance level (default = 0.05) and hypothesized differences in the means (default = 0). The *t* test two sample assuming equal variances will calculate the results (Figure 11.7).

Interpreting the t-test two-sample results

If	Then
Test statistic > critical value (i.e., $t > t_{crit}$)	Reject the null hypothesis
Test statistic < critical value (i.e., $t < t_{crit}$)	Accept the null hypothesis
$p < \alpha$	Reject the null hypothesis
$p > \alpha$	Accept the null hypothesis

FIGURE 11.7

Two-sample t test with equal variances.

	A	B	C	D	E	F
1	EDTA	PPT	t-Test: Two-Sample Assuming Equal Variances		0.05	α
2	2.98	3.69	Equal Sample Sizes			
3	2.95	3.9		EDTA	PPT	
4	2.15	3.83	Mean	3.22	3.94	
5	3.41	4.08	Variance	0.45672	0.06436	
6	3.97	3.76	Observations	6	6	
7	3.86	4.38	Pooled Variance	0.26054		
8			Hypothesized Mean Difference	0		
9			df	10		
10			t Stat	-2.443		
11			P(T<=t) one-tail	0.017	Reject Null Hypothesis	
12			T Critical one-tail	1.812		
13			P(T<=t) two-tail	0.035	Reject Null Hypothesis	
14			T Critical Two-tail	2.228		

Since the null hypothesis is that the mean difference $(x_1 - x_2)$ = 0, this is a two-sided test. Therefore, use the two-tail values for your analysis. Since the t statistic $> t_{crit}$ (2.443 > 2.228) and $p < \alpha$ (0.035 < 0.05), we can reject the null hypothesis that the means are the same. Therefore, we can say that the two anticoagulants produce a different calcium level at a 95 percent confidence level.

t Test: Two Sample Assuming *Unequal* Variances

PPT and COL have different variances, so we could use them to evaluate whether calcium levels are the same or different using these two anticoagulants:

- The null hypothesis H_0 is that the mean difference $(x_1 - x_2)$ = 0, or in other words, the means are the same.
- The alternative hypothesis Ha is that the mean difference is less than or greater than 0, or in other words, the means are not the same.

Conduct an **F** *Test to Determine if the Variances Are Equal.* Since the two recipes aren't "paired" in any way, the first step is to determine whether the variances are equal to determine which t test to use. So select the data (columns A and B in Figure 11.8), and use the QI Macros or the Data Analysis ToolPak to select the F test.

FIGURE 11.8

Two-sample *F*-test data.

	A	B	C	D	E	F	G
1	PPT	COL	F-Test Two-Sample for Variances	0.05	α		
2	3.69	3.54					
3	3.9	4.4		*PPT*	*COL*		
4	3.83	3.28	Mean	3.94	3.46		
5	4.08	2.28	Variance	0.06436	0.51		
6	3.76	3.34	Observations	6	6		
7	4.38	3.92	df	5	5		
8			F	0.13			
9			P(F<=f) one-tail	0.020	0.041	Two-tail	
10			F Critical one-tail	5.05	7.15	Two-tail	
11			One-tail	Reject Null Hypothesis because p < 0.05			
12			Two-tail	Reject Null Hypothesis because p < 0.05			

Since $p < \alpha$ (0.0041 < 0.05), we can reject the null hypothesis that the variances are equal. Now we can run the *t* test assuming *unequal* variances.

t *Test Assuming Unequal Variances.* Now use the data and the QI Macros or the Data Analysis ToolPak to select the two-sample *t* test assuming unequal variances. Enter a significance level (default = 0.05). The two-sample *t* test for unequal variances will calculate the results (Figure 11.9).

FIGURE 11.9

Two-sample *t* test for unequal variances.

	A	B	C	D	E	F
1	PPT	COL	t-Test: Two-Sample Assuming Unequal Variances	0.05	α	
2	3.69	3.54	Equal Sample Sizes			
3	3.9	4.4		*PPT*	*COL*	
4	3.83	3.28	Mean	3.94	3.46	
5	4.08	2.28	Variance	0.06436	0.50816	
6	3.76	3.34	Observations	6	6	
7	4.38	3.92	Hypothesized Mean Difference	0		
8			df	6		
9			t Stat	1.554		
10			P(T<=t) one-tail	0.086	Accept Null Hypothesis	
11			T Critical one-tail	1.943		
12			P(T<=t) two-tail	0.171	Accept Null Hypothesis	
13			T Critical Two-tail	2.447		

Interpreting the t test assuming unequal variances results

If	Then
Test statistic > critical value (i.e., $t > t_{crit}$)	Reject the null hypothesis
Test statistic < critical value (i.e., $t < t_{crit}$)	Accept the null hypothesis
$p < \alpha$	Reject the null hypothesis
$p > \alpha$	Accept the null hypothesis

Since the null hypothesis is that the mean difference $(x_1 - x_2)$ = 0, this is a two-sided test. Therefore, use the two-tail values for your analysis.

Since the t statistic $< t_{crit}$ (1.554 < 1.943) and $p > \alpha$ (0.171 > 0.05), we can accept the null hypothesis that the means are the same. The two anticoagulants produce calcium levels that are equal at a 95 percent confidence level.

t Test: Paired Two Sample for Means

A paired t test using two *paired* samples compares two *dependent* sets of test data. It helps to determine whether the means (i.e., averages) are different from each other. An example might include test results before and after training (these are paired because the same student produces two results).

The same would be true of weight loss. If a diet claims to cause more than a 10-pound weight loss over a 6-month period, you could design a test using several individuals before and after weights. The samples are "paired" by each individual. You might want to know if the diet truly delivers greater than a 10-pound weight loss. The null hypothesis is less than or equal to 10 pounds. The alternate hypothesis is greater than 10 pounds.

- $H_0 \leq 10$ pounds
- $H_a > 10$ pounds

Since the null hypothesis is stated as "less than or equal to," this is a one-sided test.

Now conduct a test with several individuals, and enter the data into Excel (columns A and B in Figure 11.10). Then use the QI Macros or the Data Analysis ToolPak to select the paired two-sample t test. At a significance level of 0.05 and hypothesized mean difference of 10 pounds, the paired two-sample t test will calculate the results.

FIGURE 11.10

Paired two-sample t test.

	A	B	C	D	E
1	Before Diet	After Diet	t-Test: Paired Two Sample for Means		
2	213.4	200.1			
3	225.0	216.4		Before Diet	After Diet
4	217.0	195.6	Mean	211.65	201.125
5	183.7	175.0	Variance	144.0107	175.1713
6	197.2	201.3	Observations	16	16
7	223.6	214.8	Pearson Correlation	0.583732	
8	224.2	215.7	Hypothesized Mean Difference	10	
9	215.2	200.7	df	15	
10	202.4	211.7	t Stat	0.181578	
11	217.7	216.1	P(T<=t) one-tail	0.429172	
12	221.0	208.5	t Critical one-tail	1.753051	
13	219.9	188.4	P(T<=t) two-tail	0.858345	
14	205.4	206.4	t Critical two-tail	2.131451	
15	195.1	180.9			
16	218.0	184.1			
17	207.6	202.3			

Interpreting the paired t-test results

If	Then
Test statistic > critical value (i.e., $t > t_{crit}$)	Reject the null hypothesis
Test statistic < critical value (i.e., $t < t_{crit}$)	Accept the null hypothesis
$p < \alpha$	Reject the null hypothesis
$p > \alpha$	Accept the null hypothesis

Since the null hypothesis is that weight loss is less than or equal to 10 pounds, this is a one-sided test. Therefore, use the one-tail values for your analysis.

Note: The two-sided values would apply if our null hypothesis were that H_0: mean difference = 10 pounds.

Since the t statistic $< t_{crit}$ (0.181578 < 1.753051) and $p > \alpha$ (0.429172 > 0.05), we can accept the null hypothesis that the weight loss is less than or equal to 10 pounds.

Example of One-Sample t *Test.* We could have cast this as a one-sample t test. If we calculate the difference between the before

FIGURE 11.11

One-sample *t* test for diet.

	A	B	C
1	Difference	t-Test 1-sample	
2	13.3	Test Mean	10
3	8.6	Confidence Level	0.95
4	21.4	N	16
5	8.7	Average	10.5
6	-4.1	Stdev	11.56526
7	8.8	SE Mean	2.891316
8	8.5	T	0.181578
9	14.5	TINV	1.753051
10	-9.3	p - One sided	0.429172
11	1.6	p - two sided	0.858345
12	12.5		
13	31.5		
14	-1.0		
15	14.2		
16	33.9		
17	5.3		

and after weights, we could test whether the difference is greater than 10 pounds (Figure 11.11).

Again, since the *p* value of 0.429172 is greater than 0.05, we accept the null hypothesis that weight loss is less than or equal to 10 pounds.

Tukey Quick Test for Means in Nonnormal Data

A *Tukey quick test* is like a *t* test, but it can handle *nonparametric* (i.e., nonnormal) data. It helps to determine whether the means are the same or different from each other. The null hypothesis H_0 is that the means are the same.

Tukey's quick test can be used when

- There are two *unpaired* samples of similar size that overlap each other. Ratio of sizes should not exceed 4:3.
- One sample contains the highest value, and the other sample contains the lowest value. One sample cannot contain both the highest and lowest values, nor can both samples have the same high or low value.

By adding the counts of the number of unmatched points on either end, one can determine the 5, 1, and 0.1 percent critical values as roughly 7, 10, and 13 points. Consider the following example.

Tukey Quick Test Example. Using data from Tukey's original paper and the Tukey quick test in the QI Macros nonparametric tools (Figure 11.12), it's easy to conduct the test. If the data violate any of the rules, the template will not calculate the Tukey quick test.

Interpreting the Tukey quick test results

If	Then
Total end count ≥ 7	Reject the null hypothesis at the 5 percent confidence level
Total end count ≥ 10	Reject the null hypothesis at the 1 percent confidence level
Total end count ≥ 13	Reject the null hypothesis at the 0.1 percent confidence level
Total end count < 7	Accept the null hypothesis

The null hypothesis H_0 is that the means are the same. In this example, since the end count = 9, we reject the null hypothesis at a 2 percent confidence level.

FIGURE 11.12

Tukey quick test.

	A	B	C	D	E
1	**Sample 1**	**Sample 2**	Sorted Combined Samples	Tukey Quick Test	
2	15	16.3	15	Total End Count	Confidence %
3	16.5	18.8	15	2	50%
4	17.3	15.8	15	3	63%
5	15.3	17.1	15.1	4	75%
6	15	17.9	15.3	5	84%
7	15.1	17.4	15.8	6	91%
8	15	16.7	16.3	7	95%
9	17.6	17.3	16.5	8	97%
10	17.4	17.5	16.7	9	98%
11	16.7	18.7	16.7	10	99%
12		19.5	17.1	End Count Table	
13			17.3	Top End Count	5.0
14			17.3	Bottom End Count	4.0
15			17.4	Total End Count	9.0
16			17.4	Signficant? (Y/N)	Yes
17			17.5	Confidence? (%)	98%
18			17.6	*p*	0.018

ANALYSIS OF VARIANCE

While t tests can handle only two samples, analysis of variance (ANOVA) can help you to determine whether two *or more* samples have the same mean or average. The null hypothesis (H_0) is that $Mean_1 = mean_2 = mean_3$. The goal is to disprove this (i.e., the samples have different means) at a certain confidence level (95 or 99 percent). Excel and the QI Macros can perform single- and two-factor analyses.

Single-Factor Analysis

From Figure 11.1, we want to compare how four different anticoagulants affect calcium levels.

Using Excel and the QI Macros, select data in columns H1:L8, and click on QI Macros "Anova and Analysis Tools—Anova Single Factor" to run a single-factor ANOVA at the 95 percent or $\alpha = 0.05$ confidence level (Figure 11.13).

Since the p value is less than α, we can reject the null hypothesis (i.e., the means are different). We could not, however, reject the null hypothesis at $\alpha = 0.01$. Just for fun, you might want to run a box-and-whisker chart on the data to see the variation (Figure 11.14).

FIGURE 11.13

ANOVA results.

	G	H	I	J	K	L	M	N	O
1	Anova: Single Factor	α	0.05						
2									
3	SUMMARY								
4	Groups	Count	Sum	Average	Variance				
5	EDTA	6	19.32	3.22	0.45672				
6	PPT	6	23.64	3.94	0.06436				
7	COL	6	20.76	3.46	0.50816				
8	AAS	6	25.32	4.22	0.32788				
9									
10									
11	ANOVA					Reject Null Hypothesis because p < 0.05			
12	Source of Variation	SS	df	MS	F	P-Value	F crit		
13	Between Groups	3.6936	3	1.2312	3.62886	0.031	3.09839		
14	Within Groups	6.7856	20	0.33928					
15									
16	Total	10.479	23						

FIGURE 11.14

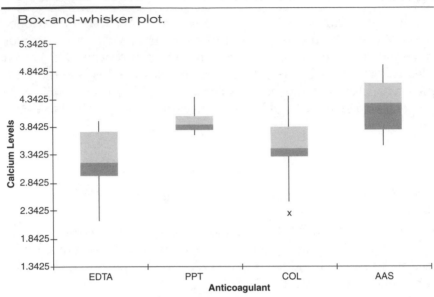

Box-and-whisker plot.

Two-Factor Analysis

To analyze a single column of data with multiple factors data, Excel requires you to set the data up in a way that can be analyzed. Figure 11.15 shows how to set up the data for two categories of patients treated with three different drugs.

Then, if you're just interested in the single-factor *drugs*, select and run a single-factor ANOVA analysis on the three drug columns (Figure 11.16).

What if you have two populations of patients (males and females) and three different kinds of medications and you want to evaluate the effectiveness of the drugs *and* the type of patient? You might run a study with two or more "replications" (more than one patient in the category receives the same drug).

Then, using Excel and the QI Macros, run a two-factor ANOVA analysis (Figure 11.17) with replication ($\alpha = 0.05$ for a 95 percent confidence).

Here, the p value for male/female is greater than α, so the means are the same. The p value for drugs is greater, so the null hypothesis holds as well (means are the same). The p value for the interaction of the drugs and patients is less than 0.05, so the effectiveness of three drugs is not the same for the two categories of patients.

FIGURE 11.15

Drug-response data.

Patient	Drug	Diffrate
Male	Drug 1	8
Male	Drug 1	4
Male	Drug 1	0
Male	Drug 2	10
Male	Drug 2	8
Male	Drug 2	6
Male	Drug 3	8
Male	Drug 3	6
Male	Drug 3	4
Female	Drug 1	14
Female	Drug 1	10
Female	Drug 1	6
Female	Drug 2	4
Female	Drug 2	2
Female	Drug 2	0
Female	Drug 3	15
Female	Drug 3	12
Female	Drug 3	9

Patients	Drug 1	Drug 2	Drug 3
Male	8	10	8
	4	8	6
	0	6	4
Female	14	4	15
	10	2	12
	6	0	9

Use Stack and Restack Tables on this data using 6 as size. Then Copy and Paste Special–Transpose.

FIGURE 11.16

Single-factor ANOVA results for drug data.

ANOVA: Single Factor

SUMMARY

Groups	Count	Sum	Average	Variance
Drug 1	6	42	7	23.6
Drug 2	6	30	5	14
Drug 3	6	54	9	16

ANOVA

Source of Variation	SS	df	MS	F	P-value	F crit
Between Groups	48	2	24	1.343284	0.290642	3.682317
Within Groups	268	15	17.86667			
Total	316	17				

ARE YOUR DATA NORMAL?

Statistical analysis may rely on your data being "normal" (i.e., bell-shaped), so how can you tell if they are really normal? The two tests most commonly used are

FIGURE 11.17

Two-factor ANOVA results for drug data.

ANOVA: Two-Factor With Replication

SUMMARY	Drug 1	Drug 2	Drug 3	Total
Male				
Count	3	3	3	9
Sum	12	24	18	54
Average	4	8	6	6
Variance	16	4	4	9
Female				
Count	3	3	3	9
Sum	30	6	36	72
Average	10	2	12	8
Variance	16	4	9	28.25
Total				
Count	6	6	6	
Sum	42	30	54	
Average	7	5	9	
Variance	23.6	14	16	

ANOVA

Source of Variation	SS	df	MS	F	P-value	F crit
Sample	18	1	18	2.037736	0.17894	4.747221
Columns	48	2	24	2.716981	0.106343	3.88529
Interaction	144	2	72	8.150943	0.00581	3.88529
Within	106	12	8.833333			
Total	316	17				

- Normal probability plot
- *p*-Value or critical-value method

Normal Probability Plot Method

If you've used any of the QI Macros X Chart templates, you know that the normal probability plot is part of the XmR, XbarR, and XbarS templates (Figure 11.18).

Just by looking at the histogram (bell-shaped) and probability plot, you can see that these data are fairly normal. The probability plot transforms the data into a normal distribution and plots them as a scatter diagram.

- Normal data will follow the trend line.
- Nonnormal data will have more points farther from the trend line.

FIGURE 11.18

Normal probability plot.

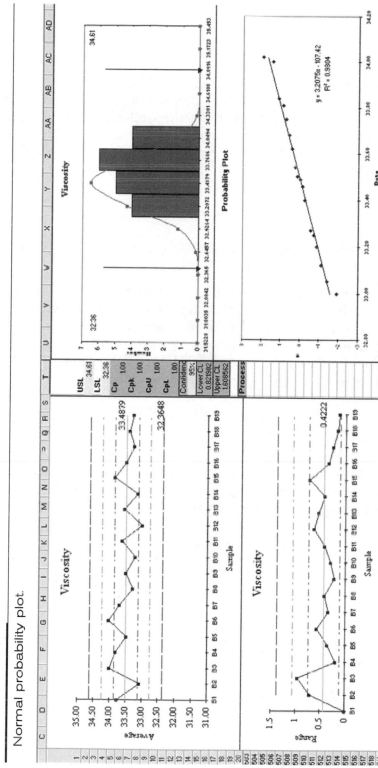

p-Value and Critical-Value Method

The descriptive statistics or normality test in the QI Macros Anova Tools uses the Anderson-Darling method to analyze normality more rigorously. The output includes the Anderson-Darling statistic, A^2, and both a *p* value and critical values for A^2. Using cells A1:A26 from the XbarR.xls in c:\qimacros\testdata, you would get Figure 11.19.

The Anderson-Darling values shown are

- $A^2 = 0198$
- $p = 0.869$
- 95 percent critical value = 0.787
- 99 percent critical value = 1.092

In this case, the null hypothesis is that the data are normal. The alternative hypothesis is that the data are nonnormal. Reject the null hypothesis (i.e., accept the alternative) when $p \leq \alpha$ or $A^2 >$ critical value.

Using the *p* value, $p = 0.869$, which is greater than α (level of significance) of 0.01. Thus we can accept the null hypothesis (i.e., the data are normal). Using the critical values, you would reject this null hypothesis (i.e., data are nonnormal) only if A^2 is greater than either

FIGURE 11.19

The *p*-value method.

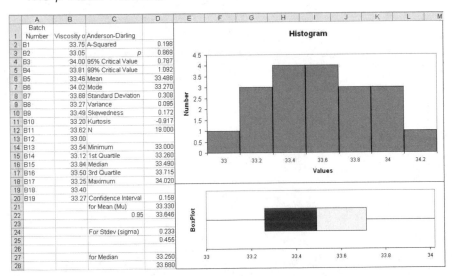

of the two critical values. Since 0.198 < 0.787 and 0.198 < 1.092, you can be at least 99 percent confident that the data are normal.

Another Example. Using cells D1:D41 (after deleting the blank row) from the XbarR.xls in c:\qimacros\testdata, you would get the result in Figure 11.20. Notice how the normality plot curves at the right so that some of the points are farther from the line. Using Anderson-Darling, we discover that the data are considered normal at one level (99 percent) but not at another (95 percent).

Using the p value, $p = 0.028$, which is greater than α of 0.01 (0.01 < 0.028 < 0.05), we can reject the null hypothesis (i.e., the data are normal) at $\alpha = 0.05$ but not at $\alpha = 0.01$. Using the critical values, since 0.787 < 0.833 < 1.092, we can reject the null hypothesis at 95 percent but not reject it at 99 percent.

Frankly, the double negative of "not rejecting the null hypothesis" makes my brain tired. All I really want to know is, "Are my data normal?" Thus, in summary:

- If the dots fit the trend line on the normal probability plot, then the data are normal.

FIGURE 11.20

Probability plot.

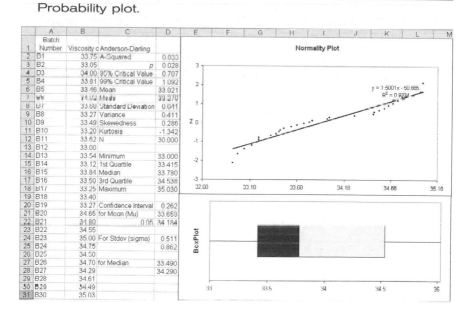

- If $p > \alpha$, then the data are normal.
- If $A2 <$ critical value, then the data are normal.

TESTS OF PROPORTION

If a manufacturer claims that his or her products are less than 3 percent defective, you can take a sample of products and determine whether or not the actual percent defective is consistent with the manufacturer's claim. Proportion testing isn't a part of Excel's Data Analysis ToolPak, but you can use the QI Macros Proportion template.

One-Proportion Test

1. If data aren't summarized, use pivot tables to summarize the trials and successes.
2. Click on the QI Macros pull-down menu, and select "Anova and Analysis Tools" and then "Proportion Tests" (Figure 11.21).
3. Enter test proportion in A3 (e.g., 3 percent is 0.03).
4. Enter number of trials in B3 (e.g., 100 products).
5. Enter number of successes in C3 (e.g., defects in 100 products).
6. Enter confidence level in E1 (e.g., 0.95 = 95 percent).
7. Null hypothesis is $H_1 = H_0$ (sample proportion = proportion). p values are calculated via direct and normal approximation methods for $H_1 = H_0$, $H_1 > H_0$, and $H_1 < H_0$.
   - If cell H3:H5 (or J3:J5) is green, you can accept the null hypothesis.
   - If cell H3:H5 is red, reject the null hypothesis.

In this example with five defects in 100 samples, you can reject the null hypothesis that $H_1 = H_0$.

FIGURE 11.21

One-sample probability test.

	A	B	C	D	E	F	G	H	I	J
1					0.95	Confidence Level				
2	Proportion	Trials	Successes	Sample p	95% Confidence Intervals		p (Direct)		p (Normal)	
3	0.3	100	5	0.050000	0.016432	0.112835	0.000	H1=H0	0.000	H1=H0
4					0.007284	0.092716	1.000	H1>H0	1.000	H1>H0
5							0.000	H1<H0	0.000	H1<H0

Two-Proportion Test

If you are sending a direct-mail piece to a group of patients, you may want to know if the proportion of patients who respond could be offered an outpatient service in two different ways. You would offer a free evaluation on half the pieces to see if you have more purchases from that group than from the other group.

Here is how to perform a test of two proportions using the QI Macros Proportion template:

1. If data aren't summarized, use pivot tables to summarize the trials and successes.
2. Click on the QI Macros pull-down menu, and select "Anova and Analysis Tools" and then "Proportion Tests." A template will open. Click on the tab labeled "Two Proportions" (Figure 11.22).
3. Enter test difference in E2. The default is 0.
4. Enter number of trials in A3 and A7.
5. Enter number of successes in B3 and B7.
6. Enter confidence level in F1.
7. Null hypothesis is $H_0: P_1 = P_2; H_1: P_1 <> P_2$.
   - If cell I3:I5 is green, you can accept the null hypothesis.
   - If cell I3:I5 is red, reject the null hypothesis.

CHI-SQUARE TESTS IN EXCEL

There are different types of chi-square tests:

- Chi-square goodness-of-fit tests
- Chi-square test of a contingency table (see later in Figure 11.24)
- Fisher exact test for 2×2 tables

FIGURE 11.22

Two-sample probability test.

	A	B	C	D	E	F	G	H	I
1	First Proportion				Difference		0.95	Confidence Level	
2	Trials	Successes	First p (P1)		0	95% Confidence Intervals		p	
3	50	10	0.2		Est. Diff	-0.086775	0.206775	0.423	P1-P2=0
4					0.06			0.212	P1-P2<0
5	Second Proportion				Z			0.788	P1-P2>0
6	Trials	Successes	Second p (P2)		0.80				
7	50	7	0.14						

Chi-Square Goodness-of-Fit Test

A chi-square goodness-of-fit test evaluates the probabilities of multiple outcomes.

Las Vegas Dice Example. Let's say that we want to know if a six-sided die is fair or unfair. We develop a null hypothesis (H_0) that the die is fair (each side will have an equal probability of coming up) and the alternate hypothesis (H_a) that one or more of the sides will come up more often:

- H_0: $p_1 = p_2 = p_3 = p_4 = p_5 = p_6 = 1/6$.
- H_a: At least one p is not equal to $1/6$.

Now test 120 rolls of the die and enter the data into Excel (cells A23:C29 in Figure 11.23). Then, in an empty cell, begin typing the formula "=chitest(." Excel will prompt for the observed and expected ranges. Use your mouse to select the observed (B24:B29) and expected ranges (C24:C29). Put a comma between the two and a parenthesis at the end, and hit "Return." The chi-square test will calculate the probability (i.e., p value) of all sides being equal.

Interpreting the chi-square goodness-of-fit results

If	Then
$p < \alpha$	Reject the null hypothesis
$p > \alpha$	Accept the null hypothesis

F I G U R E   11.23

Las Vegas dice.

	A	B	C	D
				D24 ▾ ƒ =CHITEST(B24:B29,C24:C29)
23	Las Vegas Dice	Observed	Expected	
24	1	10	20	4.55759E-05
25	2	25	20	
26	3	30	20	
27	4	20	20	
28	5	30	20	
29	6	5	20	

In these results, $p = 1.55759E\ 05$ (0.0000456), which is dramatically lower than our α value of 0.05, so we can reject the null hypothesis that the die is fair.

Chi-Square Test of a Contingency Table

A chi-square test can evaluate whether two variables are independent of each other. We've all taken surveys and probably wondered what happened. A chi-square test of a contingency table helps to identify whether there are differences between two or more demographics. Consider the following example.

Men versus Women Chi-Square Test Example. Imagine asking male and female patients if they agree, disagree, or are neutral on a given topic (e.g., nursing satisfaction). How will we know if they have the same or differing opinions? We can develop a null hypothesis (H_0) that men and women share the same views and an alternate hypothesis (H_a) that they are different:

- H_0: men = women.
- H_a: men <> women.

Now conduct the survey and enter the number of responses into Excel (cells A1:C4 in Figure 11.24). As you can see, men seem to agree more than women do, but is the result statistically different?

Select the data and use the QI Macros to select the chi-square test (this is not part of Excel's Data Analysis ToolPak). The chi-square test macro will calculate the results. Note that we don't need the same number of responses from each group to get a result.

FIGURE 11.24

Men versus women.

	A	B	C	D	E	F
1		Men	Women	Total	**Chi-Sq**	**16.16403**
2	Agree	58	35	93	***p***	**0.000309**
3	Neutral	11	25	36		
4	Disagree	10	23	33		
5	Total	79	83	162		

Interpreting the chi-square test results

If	Then
$p < \alpha$	Reject the null hypothesis
$p > \alpha$	Accept the null hypothesis

In these results, $p = 0.00031$. If the p value (0.000309) is less than the significance (e.g., $\alpha < 0.05$), we can reject the null hypothesis that men and women have the same views on the subject.

Fisher Exact Test of a 2×2 Table

A Fisher exact test evaluates small 2×2 tables better than a chi-square test because it calculates the *exact* probability. A Fisher exact test of a 2×2 table helps to identify whether there are differences between two or more demographics. Consider the following example.

Men versus Women Dieting: Fisher Exact Test Example. Imagine asking men and women whether they are dieting. How will we know if one sex diets more than the other? We can develop a null hypothesis (H_0) that men and women diet equally and an alternate hypothesis (H_a) that they are different:

- H_0: men = women.
- H_a: men <> women.

Now conduct the survey and enter the number of responses into Excel (cells A1:C3 in Figure 11.25). As you can see, men seem to diet less than the women do, but is the result statistically significant?

Use the QI Macros menu to select the "Fisher Exact Test" (not in Excel's Data Analysis ToolPak). The Fisher exact test macro will calculate the exact test statistic and the chi-square statistic.

FIGURE 11.25

Fisher exact test.

	A	B	C	D	E	F
1		Men	Women	Total	**Fishers**	
2	Dieting	1	9	10	*p 2-Tail*	**0.00275946**
3	Not Dieting	11	3	14	Chi-Sq	10.9710254
4	Total	12	12	24	*p*	0.00092527

Interpreting the Fisher exact test results

If	Then
$p < \alpha$	Reject the null hypothesis
$p > \alpha$	Accept the null hypothesis

In these results, the Fisher exact test p value is 0.00276. We can reject the null hypothesis at the 0.05 and 0.01 levels but not the 0.001 level of α.

Notice that the Fisher exact test p value is higher than the chi-square p value of 0.00093. Chi-square would let us reject the null hypothesis at the 0.001 level.

DETERMINING SAMPLE SIZES

In manufacturing applications, you often need to figure out how many samples to take to ensure that you get a valid sample size of a larger lot. From the QI Macros pull-down menu, select "Anova and Analysis Tools." Click on "Sample Size" to get Figure 11.26. Input the confidence interval and level and any other information you have to calculate the sample size required to meet your confidence needs.

To calculate a *sample size*, you need to know

1. The confidence level required (90, 95, or 99 percent)
2. The desired width of the confidence interval (±5 percent)
3. The variability of the characteristic (e.g., mean)

FIGURE 11.26

Sample size calculator.

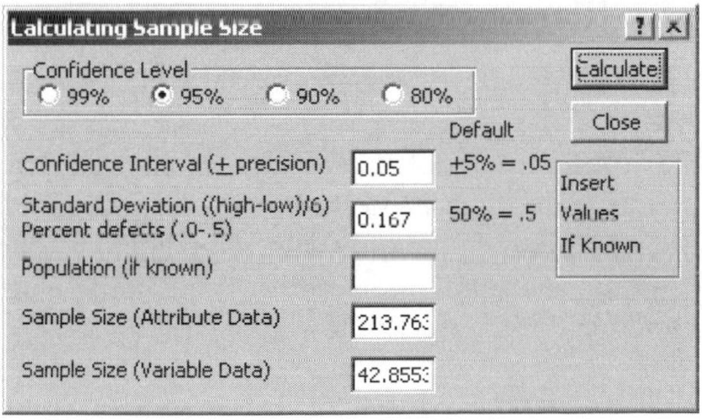

The QI Macros sample size calculator is designed to work with both variable (measured) and attribute (counted) data. The defaults are set to standard parameters:

- 95 percent confidence level
- ±5 percent (0.05) confidence interval
- Variable data with standard deviation of 0.167 (Change this value to 0.5 for attribute data.)

Confidence Level

In sampling, you want to know how well a sample reflects the total population. The 95 percent confidence level means that you can be 95 percent certain that the sample reflects the population *within* the confidence interval.

Step 1: Select a confidence level (typically 95 percent).

Confidence interval: The confidence interval represents the range of values that includes the true value of the population parameter being measured.

Step 2: Set the confidence interval (typically ±5 percent, or 0.05).

Attribute (counted) sampling: If 95 out of 100 are good and only 5 are bad, then you wouldn't need a very large sample to estimate the population. If 50 are bad and 50 are good, you'd need a much larger sample to achieve the desired confidence level. Since you don't know beforehand how many are good or bad, you have to set the attribute field to (50 percent, or 0.5).

Step 3: Attribute data—set percent defects to 0.5.

Variable (measured) data: If you know the standard deviation of your data (from past studies), then you can use the standard deviation. If you know the specification tolerance, then you can use (maximum value − minimum value)/6 as your standard deviation. (The default is $1/6 = .167$.)

Step 4: Variable data—enter standard deviation. Use the "Percent Defects/Standard Deviation" field for either attribute or variable samples.

Step 5: Enter the total population (if known).

Step 6: Press "Calculate" to read the sample size.

Use the sample size calculated for your type of data: attribute or variable. In this case, we're using variable data, so the sample size would be 43.

Regression analysis: If you think two different measurements are interrelated (i.e., there's a cause and effect), you can use regression analysis to confirm or deny that they are related. Use the scatter diagram or regression analysis tool under the QI Macros ANOVA menu to validate your suspicions.

To run regression analysis using Excel or the QI macros:

1. Select the labels and data.
2. In Excel, select "Tools—Data Analysis—Regression" or in the QI Macros select "Anova and Analysis Tools" and then "Regression Analysis."
3. Input the confidence level (e.g., 0.95)
4. Evaluate the R^2 value (>0.80 is a good fit).
5. Evaluate the F and p values.
6. Get the equation for the fitted data.
7. Use the equation to predict other values.

Door-to-Balloon versus AMI Mortality Rate Example. Imagine, for example, that a hospital wanted to know if a door-to-balloon (DTB) time less than 90 minutes affects acute myocardial infarction (AMI) mortality (Figure 11.27). Use Excel or the QI Macros to run a regression analysis at confidence level of 0.95 (Figure 11.28).

Analysis: If R^2 (0.765) is less than 0.80, as it is in this case, there is not a good fit to the data. Some statistics references recommend using the adjusted R^2 value.

FIGURE 11.27

Door-to-balloon versus AMI mortality data.

	A	B
	Patients Door to Balloon <	Acute Mycardial Infarction
1	90 minutes	mortality
2	57.1%	1.7%
3	51.5%	1.9%
4	89.5%	1.4%
5	74.2%	1.3%
6	84.6%	1.4%
7	96.0%	0.7%
8	91.7%	1.1%
9	91.3%	0.8%
10	92.6%	0.7%
11	92.4%	0.8%

FIGURE 11.28

Door-to-balloon versus AMI mortality regression.

	D	E	F	G	H	I	J	K	L
1	SUMMARY OUTPUT								
2									
3	*Regression Statistics*								
4	Multiple R	0.8746							
5	R Square	0.765	Goodness of Fit < 0.80						
6	Adjusted R Square	0.7356							
7	Standard Error	0.0022							
8	Observations	10							
9									
10	ANOVA								
11		*df*	*SS*	*MS*	*F*	*P-value*			
12	Regression	1	0.000127858	1E-04	26.042	0.001			
13	Residual	8	3.92776E-05	5E-06					
14	Total	9	0.000167136					Confidence Level	
15								0.95	0.99
16			*Coefficie*	*Standard Error*	*t Stat*	*P-value*	*Lower 95%*	*Upper 95%*	*Lower 99* *Upper 99%*
17	Intercept	0.0313	0.003880559	8.055	0.000	0.0223089	0.0402061	0.018237 0.044278
18	Patients Door to Balloon < 90 minute	-0.024	0.004649501	-5.103	0.001	-0.034449	-0.013005	-0.03933 -0.008126
19									
20	y = 0.031 -0.024*Patients Door to Balloon < 90 minutes								

Interpretation: R^2 of 0.765 means that 76.5 percent of the variation in AMIs can be explained by the percent of patients with DTB times less than 90 minutes.

Now, evaluate the p value.

If	Then
$p < \alpha$	Reject the null hypothesis
$p > \alpha$	Accept the null hypothesis

In this case, the p value of 0.001 is less than 0.05, so we can reject the null hypothesis that DTB and AMI are unrelated. This relationship also could be investigated using a scatter diagram (Figure 11.29).

Multiple Regression Analysis

The purpose of multiple regression analysis is to evaluate the effects of two or more independent variables on a single dependent variable. Select 2 to 16 columns with the dependent variable in the

FIGURE 11.29

Door-to-balloon versus AMI mortality scatter plot.

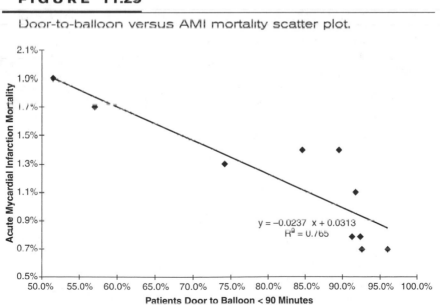

first (or last) column. Imagine, for example, that we want to know whether customer perception of quality varies with various aspects of geography and shampoo characteristics: foam, scent, color, or residue. Use Excel's Data Analysis Toolpak or the QI Macros to do multiple regression analysis on these data at the 95 percent level (Figure 11.30, matrix-plot.xls).

Evaluate the R^2 value (0.800) and F and p values

Analysis: If $R2$ is greater than 0.80, as it is in this case, there is a good fit to the data.

Then check p values:

If	Then
$p < \alpha$	Reject the null hypothesis
$p > \alpha$	Accept the null hypothesis

The null hypothesis is that there is no correlation (H_0 = no correlation). Looking at the p values for each independent variable—region, foam, and residue—we see that they are less than α (0.05),

FIGURE 11.30

Multiple regression analysis.

	A	B	C	D	E	F	G	H	I	J	K
1	Region	Foam	Scent	Color	Residue	Quality	SUMMARY OUTPUT				
2	1	6.3	5.3	4.8	3.1	91					
3	1	4.4	4.9	3.5	3.9	87	*Regression Statistics*				
4	1	3.9	5.3	4.8	4.7	82	Multiple R	0.8944731			
5	1	5.1	4.2	3.1	3.6	83	R Square	0.800082			
6	1	5.6	5.1	5.5	5.1	83	Adjusted R Square	0.7445493			
7	1	4.6	4.7	5.1	4.1	84	Standard Error	2.2055997			
8	1	4.8	4.8	4.8	3.3	90	Observations	24			
9	1	6.5	4.5	4.3	5.2	84					
10	1	8.7	4.3	3.9	2.9	97	ANOVA				
11	1	8.3	3.9	4.7	3.9	93		df	SS	MS	F
12	1	5.1	4.3	4.5	3.6	82	Regression	5	350.4359369	70.087	14.4074
13	1	3.3	5.4	4.3	3.6	84	Residual	18	87.56406307	4.8647	
14	2	5.9	5.7	7.2	4.1	87	Total	23	438		
15	2	7.7	6.6	6.7	5.6	80					
16	2	7.1	4.4	5.8	4.1	84		Coefficients	Standard Error	t Stat	P-value
17	2	5.5	5.6	5.6	4.4	84	Intercept	90.192183	4.046988933	22.286	1.5E-14
18	2	6.3	5.4	4.8	4.6	82	Region	-3.8591171	1.042809786	-3.7007	0.00164
19	2	4.3	5.5	5.5	4.1	79	Foam	1.8168936	0.35820658	5.0722	7.9E-05
20	2	4.6	4.1	4.3	3.1	81	Scent	1.0346846	0.922490779	1.1216	0.27676
21	2	3.4	5	3.4	3.4	83	Color	0.2326722	0.708123097	0.3286	0.74627
22	2	6.4	5.4	6.6	4.8	81	Residue	-4.0014884	0.811858697	-4.9288	0.00011

so we reject the null hypothesis and can say that these variables affect quality. Scent and color p values are greater than 0.05, so we accept the null hypothesis that there is no correlation.

CONCLUSION

While these tools are extremely useful for deeper analysis of your data, most Lean Six Sigma practitioners aren't ready to dive into them until they have a firm grasp of the basic measurement and improvement processes. If you want to learn more about these, consider Larry Stephens' *Advanced Statistics Demystified* (McGraw-Hill, 2004).

CHAPTER 12

Implementing Lean Six Sigma in Hospitals

The most familiar path is always the status quo.
Turnaround leaders must convince people that the
organization is truly on its deathbed.

—CHIP AND DAN HEATH

The idea of being able to truly transform the culture
within a hospital may sound like a daunting, if not
impossible, task. But achieving a genuine cultural
transformation in healthcare could be the key to solving
some of the most basic and difficult problems facing the
industry today.

—GREG STOCK, THIBODAUX REGIONAL MEDICAL CENTER

The essential methods and tools of Lean Six Sigma described in this book are easy to learn and use, but implementing Lean Six Sigma can be difficult. Over half of all implementations fail within three years. This is a one-sigma failure rate. If all you needed to do is teach everyone the methods and tools, implementation would be easy. Getting people to adopt the methods and tools is often the hardest part. There are ways, however, to ensure that the methods and tools take root and grow. Unfortunately, implementing Lean Six Sigma successfully requires an approach that runs counter to all the so-called advice available in the literature.

You've probably heard the acronym *WIIFM*—What's in it for me? It's the question everyone asks themselves when faced with any decision. Marketers constantly focus on defining the benefits of their product or service to answer this question.

I recently discovered that there's a much more insidious attitude I've labeled *IWWFM*—It won't work for me. Ask any teenager to try something new, and the answer invariably will be, "It won't work for me." But it's not just teens, it's people in business, *any kind of business*. Ask people to learn Lean Six Sigma (or anything else for that matter), and the answer is often, "It won't work for me."

EXCUSES, EXCUSES, EXCUSES

Sadly, the easiest way to avoid any kind of change is to simply proclaim, "It won't work for me," thereby avoiding having to learn or grow. To save time, maybe we should just shorten our response to this phrase by saying "IWWFM," or phonetically, "I dub-dub FM," so that it will sound as silly as it is.

The mere pronouncement of IWWFM cranks up the excuse machine: "It won't work for me because [insert lame reason here]." One of the things I've learned from research into motivation is that the mere use of the word *because* makes us all think, "Oh, he or she has got a reason it won't work." In Robert Cialdini's book, *Influence: The Psychology of Persuasion* (William Morrow, 1993), he cites a study where people standing in line for a copier were asked if someone could cut in front of them. One group was asked something like, "Can I cut in front of you?" and another group was asked, "Can I cut in front of you *because* I need to get this copied?" The first group usually refused; the second group usually allowed the person to cut in.

IWWFM is invariably followed with the word *because* and some lame excuse: "Lean Six Sigma won't work for me *because* we don't manufacture anything; we care for patients." "Lean Six Sigma won't work for me *because* we don't have customers; we have patients." It sounds so reasoned and well thought out, but it's pure poppycock.

The Blame Game

The great thing about using IWWFM is that once you've got people to buy the excuse, then you can ruthlessly blame others for your own failure to change: The competition is killing my profit margins. What do these crazy patients, families, and payers expect, perfection? My suppliers give me bad materials. My boss has unrealistic expectations of performance.

Nonsense! If you haven't tried it, you just don't know if it will work for you. Yes, you're probably overwhelmed. Isn't everyone? But you have to make time to try new things.

Peak-Performing People

Peak performers never invoke IWWFM. I read stories in the newspapers about handicapped skiers and blind golfers. People who want to do something usually can find a way to do it. Instead of mindlessly chanting IWWFM, they ask

- How will this work for me?
- What's the easiest way to try it?
- Who can help me?
- If it didn't work in the past, how can I adjust it to get the result I want?

It's so easy to blame others for our lot in life, but true courage comes from deciding that the only person holding you back is you. And the only way to stop holding yourself back is to start learning, embrace new ideas, and start making progress toward your outcomes.

If you find yourself, your family, or your coworkers muttering IWWFM, then start asking those motivating questions: "How could we adapt it to make it work for us?" "Who could help us figure out how to make it work?" "What's the easiest way to apply it?" Questions such as these overcome the mental traps of IWWFM *because.* . . . It's up to you; all you have to do is switch from IWWFM to direction-setting questions. Too easy!

Now, I know exactly what some of you are thinking: *It won't work for me because.* . . .

DECISIVE FORCE

While reading *Words that Work: It's Not What You Say, It's What People Hear* (Hyperion, 2007), by Dr. Frank Luntz, I was intrigued by his discussion of Colin Powell's doctrine of military success. Powell called the strategy of military success *decisive force* that "ends wars quickly and in the long run save lives." What Powell meant was force applied that was *precise, clean,* and *surgical.* The media misquoted him as saying *overwhelming force.* While overwhelming force sounds more exciting than decisive force, overwhelming force is about *people* and *process;* decisive force is about a *result.*

Decisive Force in Lean Six Sigma

What does Colin Powell's doctrine of military success have to do with Lean Six Sigma? Everything.

This weekend I played a round of golf with a friend from Honeywell who admitted that he'd been trained as a green belt a few years ago as part of Honeywell's train everyone in Lean Six Sigma program. He admitted that he still hadn't done a single improvement project. I'm still stunned by the number of companies that track how many belts they've trained, as if that really matters. It's like measuring the success of a war by keeping a body count.

Success with Lean Six Sigma is not how many troops you put in the field (i.e., black belts, green belts, etc.). It's about how many profit-enhancing, delay-reducing, defect-reducing, deviation-eliminating, or patient-delighting improvements you put in place and keep in place.

Another downside of this universal training program is that it takes time away from customers.

Success with Lean Six Sigma is not about how many platoons are in the field (i.e., Lean Six Sigma projects). It's about how many capture meaningful objectives. Too many teams end up majoring in minor things. They move water coolers, not mountains. Too much of the emphasis of Lean Six Sigma involves brainstorming. Letting teams choose their own projects is like turning troops loose in a country and saying, "Capture whatever you want."

Let's revisit Powell's strategy: *precise, clean,* and *surgical.* Using data about operational problems, you should be able to narrow your focus down to precise, clean problems that can be dealt with surgically by a small team of experts.

Remember the 4/50 rule:

4 percent of the steps cause 50 percent of the defects and deviation.

4 percent of the gaps between steps cause 50 percent of the delay.

Remember the dark side of the 4/50 rule:

50 percent of the effort will achieve only 4 percent of the benefit.

Overwhelming force (wall-to-wall, floor-to-ceiling Lean Six Sigma) makes it *seem* like we're doing everything possible to implement Lean Six Sigma. If it's not going well, then we get the urge to *put more troops in the field*. Training more people and starting more teams to brainstorm more problems is an invitation to disaster.

Decisive force focuses your Lean Six Sigma effort on mission-critical problems that can dramatically reduce delay, defects, and deviation while delighting customers and boosting profits.

It's up to you, but I'd like you to consider that Colin Powell's doctrine of decisive force will make a dramatic difference in the effectiveness of your Lean Six Sigma efforts. Isn't it time to start minoring in *major* things?

CRISIS JUNKIES

In most companies, including healthcare organizations, it is so much easier to fight fires and fix problems. It's so immediate and gives such a sense of accomplishment when you fix something that's a problem right now.

But it's an addiction, not a benefit. As with most addictions, the addict will do anything to get his or her next fix. In hospitals, this means resisting improvement methods such as Lean Six Sigma that eliminate the need for firefighting.

My wife works in software development. In a recent system release, her group's software worked flawlessly. Most other groups had to work around the clock for days to correct their software bugs and failures. Guess which groups got rewards for going the extra mile? You guessed it: the groups with the buggy software failures.

More often than not, reward systems fuel the failure-and-fix addiction cycle.

Rewarding crises is a mistake. Majoring in minor things such as daily problems is a mistake. But you can't just stop doing it. You will have to wean yourself off the addiction.

MINOR IN MAJOR THINGS

As I continue to train people in the simple, essential methods of Lean Six Sigma, one thing is painfully clear: Most people are too busy fighting fires to spend much time on fire prevention. They are "majoring in minor things," not "minoring in major things."

Here's my New Year's challenge to you: *Make time to minor in major things.* If you set aside two hours a week to work on prevention, mistake-proofing, and improvement, you'll make dramatic progress. Put it on your schedule, and refuse to leave the hospital until it's done.

As you reduce the daily firefighting, you'll have more time to spend on improvement. And more benefits will accrue. Productivity and profits will climb out of their rut and head for new territory.

It's not particularly hard to start making improvements. You just need a little focus:

> *Week 1:* Find or collect the data about a particularly nasty problem. Graph it over time as a control chart.
>
> *Week 2:* Use Pareto charts to narrow the cause of the problem to specific processes, machines, materials, locations, or whatever.
>
> *Week 3:* Gather the subject matter experts on the specific aspects of the problem, and analyze the root causes of the problem. Verify the causes with facts and figures. Develop countermeasures to address the specific problem.
>
> *Week 4:* Start implementing the countermeasures. Process changes take less time. Technology changes take longer.
>
> *Week 5:* Start on the next problem while monitoring the existing one.

Tools

1. *Control chart* of defects (per million opportunities)
2. *Pareto charts* of main contributors to the problem. Usually two or more levels of Pareto will be necessary to find a specific problem to solve.
3. *Ishikawa diagram* of root causes
4. *Countermeasures matrix* of potential solutions
5. *Results graphs* (line, Pareto) of improvements

If you go to www.qimacros.com/hospitalbook.html, you can download the QI Macros Lean Six Sigma SPC Software 90-day trial. Use the Control Chart Wizard to draw control charts. Use the Pareto chart macro to draw Pareto charts. Click on "Fill-in-the-Blank Templates" to access the Ishikawa diagram.

Nothing will get better until we carve out time to find and prevent the root causes of mistakes, errors, defects, delay, and deviation. How we choose to spend our time will determine our progress at the end of the year.

Do you want the end of this year to feel the same way last year did? Or would you rather spend a little time on mission-critical improvement projects? Are you going to major in minor things this year or minor in major ones? The choice is up to you.

GETTING THE RIGHT PEOPLE INVOLVED WITH LEAN SIX SIGMA

Have you ever noticed that some people just seem to be mentally *wired* for improvement work, whereas others just can't seem to get it? Is there a mindset that's fertile ground for Lean Six Sigma and others that are not? Is there a way to determine this in advance?

Based on my 20-year research into the science of mindsets and motivation, I believe that the answer to these questions is yes. There is a mindset that is fully prepared to embrace the methods and tools of Lean Six Sigma, and there are ways to detect it.

THE IMPROVEMENT MINDSET

In 1990, I started learning the science of neurolinguistic programming (NLP). I think of NLP as software for your mind. With NLP, you can discover how people think and what makes them successful, as well as what makes them fail. One of the most interesting discoveries from NLP involves intrinsic motivation.

In my book, *How to Motivate Everyone* (LifeStar, 2002), I explore the five core motivation traits and two or more conflicting points of view. A few of them are key to understanding the improvement mindset.

The Improver Mindset

Ask yourself this question: *What's the relationship between your job or work this year and last year?* There are three answers to this question that reveal a lot about whether you have the right mindset for improvement:

- It's pretty much the *same.*
- It's *improved*, enhanced, expanded, and enriched.
- What do you mean *relationship?* There is no relationship. Are you asking me what's *different* about my job?

People who answer, "It's pretty much the same," want things to stay the same. *You can't stay the same and improve.* (These *traditionalists* represent 5 percent of the population. They'd rather kill Lean Six Sigma than try it.)

People who answer, "What do you mean *relationship?*" want everything to be *different.* You can't do things differently all the time and improve. Given a foolproof method of making a million dollars, these people have to fiddle with the formula *before* they try it. (These *revolutionaries* represent 30 percent of the population. They'd rather invent their own improvement method than try Lean Six Sigma.)

Like Goldilocks and the Three Bears, the middle answer is just right. (Improvement-oriented *evolutionaries* represent two-thirds of the population, but they have to spend a lot of time wrestling the traditionalists and the revolutionaries.)

If you're interviewing employees for black belt or green belt training, ask them,

What's the relationship between your job or work this year and last year? Then listen to their answers. If they don't answer *better, improved, expanded, enhanced,* or *something similar,* it might be time to look for someone else.

Process-Oriented Mindset

Lean Six Sigma focuses on processes and systems, so to be good at Lean Six Sigma, you've got to like processes. Ask yourself, *Why did I choose my current job or work?* There are two answers to this question:

- People who answer with a *list* of criteria (e.g., a chance to learn, meet new people, make more money) have an *innovator* mindset.
- People who feel confused and don't know how to respond feel that they didn't choose their job; their job chose them.

They feel compelled to tell a *story* about how their job chose them. These people have a *process* mindset.

Innovators, like revolutionaries, aren't interested in improving the existing process. They want to create a new one. Processors, on the other hand, like process. They are wired for process. They live for process.

Problem-Solver Mindset

Lean Six Sigma is a problem-solving process. Do you like to solve problems? Ask yourself two questions:

- *What's important about my job or work?* (You often get answers that correspond to one of these categories: people–relating, places–being, activities–doing, knowledge–learning, or things–getting/having.)
- Pick one of the answers (e.g., learning) and ask yourself, *Why is this (e.g., learning) important?*

Again, there are two answers to this question:

- If your answer describes what you'll accomplish or *achieve*, then you're an *achiever*: "I want to get a black belt certification to advance my career and increase my earnings."
- If your answer describes the pain and suffering you'll *avoid*, then you're a *problem solver*: "I can avoid the mistakes most people make because they aren't prepared." Problem solvers often speak in *"not* language": "Using Lean Six Sigma means that our customers *won't* have to put up with all the problems normally associated with our diagnostic and treatment processes."

Lean Six Sigma is inherently a problem-solving process; it is not a goal-achievement process. If you want to be good at Lean Six Sigma, you'll want to be a problem solver.

If you're interviewing employees for black belt or green belt training, ask them,

What's important about your work? and Why is that important?
Then listen for *"not* language" and the avoidance of consequences, not achievement. You want people who love to solve and avoid problems.

Leader Mindset

Implementing Lean Six Sigma requires leadership on multiple levels. Are you a leader? Ask yourself this question, *How do I know that I have done a good job?* Again, there are two answers to this question:

- "People tell me."
- "I just know." (Often these folks will touch the middle of their chest when they say this.)

The first answer comes from a *follower*. They can follow, but they need leadership. The second answer comes from a *leader*, who can take in information and decide for himself or herself.

Lean Six Sigma belts need to be leaders. If you're interviewing employees for black and green belt positions, ask them, *How do you know that you've done a good job?* If they only say, "People tell me," keep looking. If they say, "I just know," you're on the right track.

The Lean Six Sigma Mindset

There you have it. There is a mindset that is ready for Lean Six Sigma. And two out of every three people have some of it ("evolutionaries"). Out of those two, you'll have to ask some questions to find the rest of the recipe for success.

The people who are wired for improvement have the motivation traits of *leaders*, *problem solvers*, *processors*, and *evolutionaries*. The people who are wired for innovation have the motivation traits of *achievers*, *innovators*, and *revolutionaries*.

Every business needs innovators and improvers. You just want to get them into the right jobs. Innovators help to create the future. Improvers ensure that you'll maximize your profits and minimize your risks.

The Motivation Profile

If you're still uncertain about how to ask these questions and evaluate the answers, you can take my online Lean Six Sigma Profile at www.qimacros.com/profile/six-sigma-mind-set.pl or the complete personality inventory at www.qimacros.com/nlpstyle.html.

Once you've got the right mindset, learning Lean Six Sigma requires the right kind of practice.

TALENT VERSUS PROCESS

Once you've learned the methods and tools of Lean Six Sigma, you'll need to practice the skills to make them part of your mind-set and toolkit. In his book, *Talent Is Overrated* (Portfolio, 2008), Geoff Colvin argues that talent, IQ, or "smarts" aren't what make people successful; it's what he calls *deliberate practice*. In *Outliers: The Story of Success* (Little, Brown and Company, 2008), Malcolm Gladwell argues that it can take 10,000 hours of practice (Colvin says 10 years) to achieve mastery in your chosen field. I'd like you to consider that hospitals are no different.

If you go to a golf driving range, you'll see lots of people hitting golf balls, but very few are practicing *deliberately* in ways designed to improve their performance. Tiger Woods will step on a ball in a sand trap to practice getting the ball out of a plugged lie. And he'll keep doing it until he's mastered the shot. That's deliberate practice.

What Is Deliberate Practice?

Deliberate practice

- Is designed to improve performance.
- Can be repeated *a lot*.
- Provides continuous feedback on results.
- Is highly demanding mentally.
- Isn't much fun.

Frankly, most people would rather come to work and mindlessly do the same old thing, and that's what a lot of people do. They do the same thing over and over again without questioning the whys and hows of doing it.

Deliberate Practice and Lean Six Sigma

Colvin says, "Opportunities to practice business skills directly are far more available than we usually realize." Lean and Six Sigma are precise forms of deliberate practice. Colvin found that an average tennis player and a pro or a good worker and a great worker have many differences:

- *Great players understand the significance of indicators that average performers don't even notice.* Using control charts, histograms, and Pareto charts, great companies can detect tiny shifts in process performance that are *invisible to the naked eye.*
- *They look farther ahead.* Using the voice of the customer, Quality Function Deployment (QFD), Failure Modes and Effects Analysis (FMEA), and so on, great companies look into the future of what customers really want and how to deliver those wants with a minimum number of mistakes, errors, or problems.
- *They make finer discriminations than average performers.* While most companies react to defect rates higher than 1 percent (10,000 parts per million), great companies react to defect rates greater than 3 parts per million.

Applying Deliberate Practice

When a problem occurs, good companies fix the product or service, but great companies go back and *fix the process that created the problem.* When creating a new product or service, good companies bootstrap the product, but great companies *design for Six Sigma.* Good employees like to fight fires; great employees like to prevent fires.

Here's my point: Are you going to spend part of every day in deliberate practice, getting better, faster, and cheaper? Are you committed to going from good to great? Or are you comfortable being average? It's up to you.

NEW CEOS CAN KILL LEAN SIX SIGMA

I just got an e-mail from a healthcare master black belt whom I greatly respect:

> After months of waiting on a decision from our new CEO, we were informed that our last day is near, but we have opportunity to apply for other jobs internally. The new CEO is not a supporter of Lean Six Sigma. We have saved over $20M in hard-dollar savings, trained hundreds who have advanced their careers, and made major improvements. It is very disappointing, but after multiple CEOs in a few years, it was going to happen.

The former CEOs have all been great supporters, so the new guy is going to do things differently. He uses "What works." No one knows what that is yet.

This shows one of the great fallacies of the "get top-level commitment for Lean Six Sigma." When there's a change in leadership, Lean Six Sigma can go away, even if you have real savings.

According to executive recruiting firm Spenser Stuart, in 1980, CEOs served an average of eight years. By 2005, that number was seven, and it dropped to five years for the Fortune 500. It's closer to three years now. It may even be worse in healthcare.

BE A MONEY BELT

Keep a record of your projects and the savings associated with them. If you can't measure the dollar savings, you're on the chopping block. If you measure them, then you have leverage with the new leadership and a great résumé even if the new leader doesn't care.

With the economy in trouble, every business leader is going to be looking at the bottom line—what contributes and what doesn't. The quality department always has been an easy target.

It would be great if we could convert every business to a quality management system, but a new CEO isn't brought in to keep things the same; he or she is brought in to shake things up. And many times the good is thrown out with the bad.

Lean Six Sigma is one of the things that new CEOs kill in favor of innovation. When innovation fails, the next CEO will bring in some form of improvement methodology, maybe the next iteration of Lean Six Sigma.

I believe that the best we can hope for is to create some "money belts" who can find and fix problems that contribute to lost profits. CEOs may come and go, but money belts are invaluable and can work their magic in any climate or culture.

If the corporate culture matures to the point that it embraces a quality management system or some customers demand such a system as a condition of doing business, fantastic. Until then, be a money belt! Tell everyone stories of your improvements. Make yourself invaluable to the business. Then, even if the quality department goes away, you and the improvement methods will remain.

Learn how to be a money belt. Go to www.qimacros.com/moneybelt.html to take my free online Lean Six Sigma money belt training.

OUR REWARD SYSTEMS ARE BROKEN

One of the biggest challenges to Lean Six Sigma is existing reward systems. They are rewarding the wrong things.

Rewarding the Fix-It Factory

When I was in the phone company, thousands of people were fixing incorrect orders, bills, returned mail, and installation and repair errors. Every one of these employees got raises and bonuses based on fixing errors, not preventing them.

According to one *BusinessWeek* article I read, 15 out of every 100 patients are misdiagnosed. Doctors still get paid for the visit. I've been in emergency departments (EDs) when patients return after being discharged because they still have undiagnosed symptoms. This is rework!

Rewarding Training Not Results

Too many Six Sigma implementations measure and reward the quality department for the number of people trained, belts certified, and teams started. Let's face it, it doesn't matter how many people you train if they don't generate bottom-line, profit-enhancing results.

Reward Systems Encourage the Status Quo

Existing reward systems encourage managers and employees to maintain and enhance the status quo. Even though President Obama was elected on a platform of change, everyone wants someone else to have to change. "It's not me; it's the other guys." "If only my suppliers would change, I could do a better job." "If only my patients would change, I could do a better job." "It's not my fault."

I'd like you to consider that existing reward systems are holding you and your hospital firmly in the grip of delays, defects, and deviation that cost you patients and profits. Without improvements to the rewards system, no one will be motivated to fix the process problems that plague the company. Your employees will just keep plodding along, fixing the products and services that are broken. They'll lament over coffee breaks about how screwed up it is.

Everyone is afraid that if they use Lean to be faster that there won't be enough work and someone will get laid off. Fix-it factories worry that if the production line stops making defects, they won't have a job.

Get over it! If a hospital accelerates the delivery of quality products and services, patients will flock to it. Patients and their families don't have time to deal will slow or crappy care.

Reward systems are one of the major barriers to implementing the kinds of improvements possible with Lean Six Sigma. They are also one of the hardest nuts to crack in any business because everyone is afraid of how change will affect them. However, if you want to accelerate the benefits of Lean Six Sigma, you'll have to remove this roadblock and pave the path to endless improvement. Otherwise you're doomed to slide back into sluggish, defective ways of mediocre performance. It's up to you.

BARRIERS TO LEAN SIX SIGMA

At the 2009 American Society for Quality (ASQ) World Conference, Joe De Feo, president of the Juran Institute, gave an interactive session on the future of quality. Like everyone else, I probably heard only what I wanted to hear, but the message was clear: The future is upon us, and it will be different from the past.

While the past was predominantly about reducing manufacturing variation, the present moment shows signs that quality will be about leaner, greener, global real-time information systems. To improve quality now, there will be massive numbers of transactions—"information that has to be decoded."

Transactional Lean Six Sigma

Where transactions used to happen more slowly, they now happen in real time. What are real-time information systems? On Memorial Day, I had two people in Poland order the QI Macros software, pay with a credit card, and download the software. Electronic medical records keep track of a patient's treatment down to the last medication he or she received. A laptop purchased at a computer store in Denver triggers a supply-chain event that initiates a new laptop's construction at a factory in China.

What does this mean? While reducing variation in manufacturing is still important, the service side of the business, especially the information systems that power these customer and patient interactions, has become critical to quality.

Lower the Barriers to Quality

One participant in the ASQ session suggested that we need to "lower the barriers of entry to quality." Joe De Feo offered some key thoughts on how to lower those barriers:

- Low-cost quality management ("money belts")
- Standard work practices (e.g., checklists)
- Real-time analysis
- Scorecards and dashboards (www.qimacros.com/dashboard-scorecard-for-excel.html)
- Individual process improvements

While Lean Six Sigma has been focused on high-cost training, software, teams, and variation, the future belongs to low-cost just-in-time training, software, individuals, and information transactions.

For the last several decades, I've sought to lower the barriers to quality with low-cost software such as the QI Macros and my Excel-based scorecards and dashboards. I've applied Lean Six Sigma to itself to identify the essential methods and tools of quality, which led to the *Six Sigma Simplified* (LifeStar, 2004) and *Lean Six Sigma Demystified* (McGraw-Hill, 2007) books. I've used my software background to identify easy ways to analyze the flurry of transactions produced by most information systems (www.qimacros.com/pdf/dirty30.pdf). And I've championed the idea of "money belts"—employees who can find and fix the problems of unnecessary delay, defects, and deviation.

HERE'S MY POINT

To implement Lean Six Sigma, most companies spend a lot of money developing a hierarchy of green and black belts. Lately, I've heard customers complaining that even the best-trained black belts don't seem to know how to plug the leaks in cash flow caused by poor quality. I've heard top consultants say that companies would be better off with a few green belts who get coached through their initial projects by "money belts."

By making Lean Six Sigma sound complex and expensive, we've discouraged too many businesses from learning the essential methods and tools of Lean Six Sigma. We've stopped them from making improvements in their mission-critical processes. With our focus on variation, we've discouraged information technology (IT) departments from considering Lean Six Sigma. The hierarchy of master black belts, black belts, and green belts has created a problem of the haves versus the have-nots. Isn't it time we got over our arrogance about Lean Six Sigma and started lowering the barriers to entry for everyone?

Over the years, I've written many articles on this subject, backed by research into how companies adopt, adapt, or reject change. The answers are out there, and they are surprisingly simple, but to apply them, we will need to challenge conventional wisdom and act in the face of ridicule.

MANAGEMENT BY QUALITY (MBQ)

I just attended the Worldwide Conventions and Business Forums (WCBF) Six Sigma Summit in Chicago. The overriding theme of the conference from presentations by Forrest Breyfogle, Thomas Pyzdek, and Peter Pande was that American business leaders need MBQ to avoid the kinds of economic disasters we've seen lately. Management schools and MBAs were all skewered on the facts of their failure. Each speaker issued a challenge to the quality community to elevate Lean Six Sigma methods and tools to better manage American business.

Changing the Course of the *Titanic*

While saying leaders should use MBQ sounds good, it's about as likely to happen as the moon reversing its orbit. Consider how much *unlearning* would be required to make room for MBQ. Consider all the leaders and managers who have succeeded by gut feel, trial and error, and common sense. They aren't going to suddenly switch to a completely new management method without a significant reason to do so. An economic crisis isn't going to do it. We have to think of a better way.

Lean Six Sigma Isn't Blameless

The quality community is not without blame. As most speakers noted, Lean Six Sigma has become a project-oriented methodology with little or no focus on the long-term evolution of businesses and management. Peter Pande also pointed out that a narrow focus on process can prevent a company from seeing opportunities. We have become too narrowly focused on improvement projects. We need to think bigger.

Stealth Conversion

Let's face it, most quality practitioners aren't in a position to convert leaders to MBQ. We aren't part of the C club. But there are a couple of "stealth" ways to go about it.

1. The July–August 2009 *Harvard Business Review* (pp. 90–91) had an article entitled, "Shareholders First? Not So Fast . . . ," by Jeffrey Pfeffer, that argued against "shareholder capitalism" in favor of "stakeholder capitalism" (i.e., "customers, employees, suppliers, shareholders, and the culture at large"). This is the kind of article quality professionals can highlight and send to their leadership team to start shifting the business toward MBQ. The *Harvard Business Review* can tell leaders things that quality professionals cannot.

2. When I worked in the phone company IT department, we had several legacy information systems that were growing too cumbersome and expensive to maintain. Several attempts were made to rewrite these monoliths at an expense of hundreds of millions of dollars. Unfortunately, systems that have been around for two decades have too much embedded wisdom; it was impossible for any team to get their minds around all the requirements. And the business was changing (e.g., cell phones, fiberoptics, cable, etc.). In the time it would take to rewrite the system, the business would have changed too much for the new system to work.

   I was responsible for maintaining several smaller legacy systems. I used a different approach. Whenever I went into a program or module to make a change, I would edit or rewrite the code to optimize future maintainability.

Little by little, line by line, I bought these dinosaurs more time and saved several from extinction.

I'd like to suggest that we can do the same with quality.

Shortcut to MBQ

Anyone who thinks that he or she can institute a whole new way of running a business is probably crazy. There are too many forces holding the old one in place. Remember *reengineering*? Far too many efforts crashed and burned, almost taking companies down with them.

I'd like to suggest that every Lean Six Sigma project can achieve its project goals and add a little dose of MBQ. Maybe the control phase initiates a dashboard of management metrics to help managers keep a finger on the pulse of operations. Maybe a dash of systems thinking that links up cross-functional work. Maybe add a pull system with some one-piece flow.

Little by little, project by project, Lean Six Sigma can add MBQ to any business system and straighten out the value stream. With every project Lean Six Sigma can steer a business away from the icebergs in its path and stealthily convert managers and leaders to a new system of management.

Chaos theory says that a butterfly flapping its wings in Brazil can raise a tornado in Texas. I'd like to invite you to be the butterfly's wings.

Teaching MBQ

I had an interesting discussion with Forrest Breyfogle, who has created his Integrated Enterprise Excellence system for management. Forrest is a statistician to the marrow of his bones. He would like every manager and leader to have his level of understanding of variation. Heck, I'd like to have his understanding of variation, but he and I disagree on how to get managers to that level of understanding. Imagine an S-curve of statistical thinking (Figure 12.1).

While it might be fantastic if every manager or leader (or employee, for that matter) could suddenly arrive at *statistician-level understanding*, it's unrealistic to expect it. If, however, we borrow the perspective of marketers who use this curve to invite people to join at whatever level they are comfortable and then lead them up along the curve, I believe we can succeed without alienating anyone.

FIGURE 12.1

Lean Six Sigma learning curve.

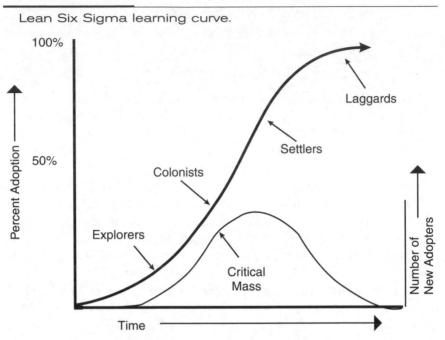

It's a covert, stealthy approach to shifting the business management system. Weave the methods and tools of Lean Six Sigma into the fabric of business operations with every project you do. Little by little, employees, managers, and leaders will learn the methods and tools of MBQ. Little by little, the business will shift to a more robust way of measuring and delivering quality on time and under budget—which will delight customers, which will delight shareholders, which will deliver value to society.

I wish we could just flip a switch to change cultures and management systems, but that's just not likely to succeed. Instead, we should be like the waves that come back again and again, wearing down the rocks that stand in our way.

WHY DO SIX SIGMA TEAMS FAIL?

I worry when I hear about companies shutting down their Six Sigma effort. What's the problem? Six Sigma should be helping, not hurting.

Years ago, when I first got started with improvement methods, I used a top-down, CEO-driven, all-or-nothing approach to implementation, just like companies are trying to do today. Following the guidance of million-dollar consultants, I started and trained hundreds of teams that met for one hour a week. Two years later, only a handful of teams had successfully solved a key business problem. Most were mired in the early steps of the problem-solving process.

So I decided to try something radical: I applied the improvement method to the improvement method. I looked at

- Each stuck team as a "defect"
- The "delays" built into process:
  - The delays between training and application
  - The delays between team meetings

Solution: Use the 4/50 rule to laser-focus improvement efforts for maximum benefit.

Root-Cause Analysis

I took a step back and started asking, Why? Why? Why? Why? Why? Here's what I found:

- *Teams are formed before the data are analyzed.* Why is this a problem? Unfortunately, when you bring a group of people together but don't have a clear focus driven by data, you end up with what I call the "100-yard dash for the directionally impaired."

Solution: Analyze the data first to narrow your focus using control charts and Pareto charts; then pick a team that has expertise in that area to do the root-cause analysis.

- *Teams choose their own problem to work on.* Why is this a problem? Unfortunately, most of the time, teams want to fix their suppliers or their customers. Union employees want to fix management; management employees want to fix workers.

  Or they choose something trivial to "get experience." Remember the dark side of the 80/20 rule: 80 percent of the effort produces only 20 percent of the benefit. This is why Six Sigma fails.

Solution: Let the data lead you to a problem that you can solve and that you own. You can't fix someone else's problem. You can give them the data and the analysis, but you cannot solve it for them. They won't implement your solution.

- *Define, Measure, Analyze, Improve, Control (DMAIC).* Why is the Six Sigma improvement process a problem? Unfortunately, DMAIC begins with *define* and *measure*, so most teams get lost in defining the process and implementing new measures.

Solution: Skip define and measure; go straight to analyze, improve, and control. You already have enough data. Somebody somewhere is keeping a count of the number and types of mistakes, errors, defects, repair, rework, or waste. Find some data you can analyze to narrow your focus. Then improve and control.

- *Chartered teams meet one hour a week forever.* Why is this a problem? Because there's 167 hours of delay between meetings. It violates the Lean principles of eliminating delay and one-piece flow.

Solution: SWAT teams. When the data have been analyzed and the problem-solving effort has been laser-focused, a team of subject matter experts (SMEs) only needs to meet for two to four hours to identify the root causes, countermeasures, and an implementation plan. Using root-cause SWAT teams, I was able to eliminate the delay between team meetings. Solutions that used to take months now took only hours.

- *Scope creep.* Teams invariably want to solve world hunger, boil the ocean, and fix everything all at once. When teams scatter their focus, they solve nothing.

Solution: Use Pareto charts to narrow the focus. Then analyze one "big bar" of the Pareto chart at a time.

- *Whale-bone diagramming.* If a team starts covering the conference room walls with fishbone after fishbone diagram, the focus is more like a flashlight than a laser.

Solution: Go back to the data and narrow the focus.

- *"Just in case" training.* Many teams and team leaders (e.g., green or black belts) get lots of training long before they apply it. Why is this a problem? Because humans lose 90 percent of what they learn in 72 hours if they don't apply it immediately.

  Looking at this from a value-added flow perspective, the delay between training and application isn't just about the waste of time but also about loss of skill. The only way to reengineer this problem is to eliminate the delay—just-in-time training.

  In the early 1990s, when I was lured into the in-depth training paradigm, I'd spend a week using a Deming prize–winning methodology to train 20 team leaders. They, in turn, would start teams that met once a week for an hour. Months went by. Years went by. Nothing got better. Here's what I discovered: You can't learn to swim without getting wet.

  So, unbeknownst to my company leadership, I changed the process. I shortened the training down to a couple of introductory hours that I would teach only immediately prior to solving a real problem. Then, in a day or two, I'd guide the team to a solution. They got experience and the good feelings associated with success. Surprisingly, many of these team members then could apply the same tools and process to other problems with equal success. I discovered that I was creating highly skilled but essentially untrained team leaders in a matter of one day. To strengthen their abilities, I'd occasionally conduct a one-day intensive session to review what they'd learned through experience. This helped to reinforce what they knew and fill in any gaps.

  With one day of experience and a day of review training, I was accomplishing what the old week of training and endless meetings could not. And we were getting bottom-line benefits simultaneously.

  Sadly enough, by the time I figured this out, the quality department was on its last legs because it had failed to do more than waste time and money defining and measuring cumbersome, error-prone processes that needed major repair. A year later, the department was disbanded and the people laid off.

Don't let this happen to you. Consider using just-in-time training to prep your teams for immediate immersion in problem solving or statistical process control (SPC). Use real data. Use real problems. From the time we are born, we learn by watching other people do things. When you guide a team through the process, team members learn an enormous amount just by watching you. Then reinforce what they've learned unconsciously with one-day review training.

You'll save your company time and money, get immediate results, encourage the adoption of Lean Six Sigma by satisfied employees, generate good buzz, and have more fun.

Solution: Using just-in-time (JIT) training, I was able to close the gap between learning and application. Give the SWAT team an hour of training and then throw them right into root-cause analysis. They'll learn more working on a real problem than they'll ever learn in training.

■ *Poor integration of Lean Six Sigma into the business.*

Solution: Using the power of "diffusion," I was able to weave the methods and tools of Lean Six Sigma into the organization with a minimum involvement of key resources. By systematically applying the improvement process to itself, I found ways to eliminate the failures and accelerate the delivery of results.

If Lean Six Sigma isn't producing immediate, measurable, ongoing benefits, management will kill it. Teams can't afford to waste time. Mistakes in team formation, team meetings, and data analysis can doom a team's chances of success. And team failures can kill Six Sigma.

"BRIDEZILLA" MEETS LEAN SIX SIGMA

Our daughter, Kelly, got married in September. Our oldest grandson, Jake, had his coming-of-age party in July. One of the themes I've discovered in these ceremonies is the tendency to think that the lavishness of the event determines its quality. My wife and I, on the other hand, got married in our backyard. Our marriage has lasted

22 years and seems to be getting better every year. What has this got to do with Lean Six Sigma? Everything.

Far too many people judge Lean Six Sigma success by how much money is thrown at it. When Six Sigma becomes an event rather than a process, it is doomed.

Ask yourself:

- How long does Lean Six Sigma stick in an organization (three years is the average tenure of a CEO)?
- What are the costs of Lean Six Sigma?
- What are the bottom-line benefits of Lean Six Sigma?
- Is Lean Six Sigma an event like a wedding or a process like a marriage?

HERE'S MY POINT

Is your Lean Six Sigma process like a celebrity wedding or a life-long commitment to excellence? Are you a black belt, green belt, or "money belt"? Is Lean Six Sigma paying for itself or just a ceremony everyone attends before they return to their "real" life?

Start using these insights gleaned from two decades of process improvement to make sure that Lean Six Sigma delivers on its promises or start looking for a new job. Once you learn the Lean Six Sigma mindset, methods, and tools, you can use them anywhere. Start now to make your company wildly successful, and you will be too.

MAKING LEAN SIX SIGMA SUCCESSFUL

> People follow tradition. This might be acceptable in private life, but in industry, outdated customs must be eliminated.
>
> —TAIICHI OHNO

Even though every leader claims to understand the 80/20 rule, many still try to deploy improvement methods everywhere. But Lean Six Sigma is like peanut butter—the wider you spread it, the thinner it gets. Remember the dark side of the 80/20 rule: If you try to use Lean Six Sigma everywhere, 80 percent of your effort will produce only 20 percent of the benefit.

In the 1990s, our CEO committed to quality. Millions of dollars and almost five years later, the company abandoned total quality management (TQM). Having the CEO on your side may help, but it's not the Holy Grail of gaining organization-wide commitment to quality.

If you've read anything about quality improvement, you've seen it repeated endlessly that you want to get top-leadership commitment. The emerging science of complexity suggests that this is a mistake. Getting CEO commitment invokes what complexity scientists call the *Stalinist paradox*, which lowers your chances of success to 50:50. The emerging science of networks suggests that it's *never* the formal leadership that determines the success or failure of a culture change, it's the *informal* leaders—the hubs—in any "network" that determine success. Informal networks are more like spider webs or wagon wheels, not hierarchies.

Formal Network versus Informal Network

In *The Tipping Point* (Little Brown, 2002), Malcolm Gladwell argues that any idea "tips" into the mainstream when sponsored by one of three informal leaders: connectors, mavens, or salespeople. Seth Godin calls these people "sneezers" because they sneeze the "idea virus" in ways that get it into the minds of everyone. Godin separates sneezers into two categories: the powerful and the promiscuous. Powerful sneezers do it because it enhances their status. Promiscuous sneezers (i.e., salespeople) do it for the money.

- *Connectors connect people with other people they know.* Think about your own company. Who is the center of influence who knows everybody and introduces everyone to everyone else?
- *Mavens connect people with new ideas.* Who is the center of influence in your company who gets everyone on board with all the new changes in technology (e.g., Lean Six Sigma, SPC, etc.)? I think of myself as a maven because I'm trying to connect you with the powerful ideas in Lean Six Sigma.
- *Salespeople do it for money.* When you follow the CEO commitment rule, these folks will show up like vultures to a carcass. Beware.

Don't Confuse the Means with the Ends

To increase results, narrow your focus. Too many hospitals are losing sight of the objective when it comes to Lean Six Sigma. The goal is to cut costs, boost profits, improve patient outcomes, and accelerate productivity; it is not the wholesale implementation of an improvement methodology.

At the ASQ's annual conference, many people stopped by my booth, drawn by the promise of *Lean Six Sigma Simplified*. They'd been buried in an avalanche of conventional folklore that you have to make a major commitment, spend lots of money training team leaders, and wait years for results. Every one of these disheartened business owners voiced the same questions: "Isn't there a better way?"

Of course there is, because all the conventional wisdom and hype about improvement methods such as Lean Six Sigma are *dead wrong!* The goal is bottom-line profit-enhancing, productivity-boosting results! Lean Six Sigma is merely a means to that end, nothing more. It is not the one-size-fits-all universal cure to what ails your business. Lean Six Sigma is a power toolkit for solving three key business problems:

- *Delay*—when the customer's order is idle
- *Defects*—errors, mistakes, scratches, imperfections
- *Deviation*—when the process, machines, or materials vary

Linear versus Circular Causes

Lean Six Sigma works very well on problems with linear cause-effects. If you step on the gas pedal in your car, for example, the car accelerates. This is a linear cause-effect. Lean Six Sigma doesn't work well on problems with *circular or systemic cause-effects*. In other words, you can't use Lean Six Sigma *directly* to change morale or customer satisfaction. If you engage employees in improving the business, morale may improve. If you improve your products and services, customer satisfaction may improve, but you can't improve morale or customer satisfaction directly. With Lean Six Sigma, you can directly engage the power laws of speed and quality.

Bell-Shaped Mindset

Because quality principles evolved predominantly in a manufacturing environment, there's a lot of emphasis on variation, shown as the *normal* or *bell-shaped* curve, where product measurements are distributed across a range of values. Unfortunately, this emphasis has blinded most leaders to the reality that defects tend to cluster in small parts of the business; they aren't spread all over like butter on bread.

What if you could get over half the "benefit" from Lean Six Sigma by investing in just 4 percent of the business? You can! Pareto's 80/20 rule is a power law. Power laws aren't linear; they grow exponentially. Therefore, if you believe in Pareto's rule, you have to believe that it applies within the 20 percent. Four percent of the business will cause 64 percent of the waste and rework. Wherever I go, I find that 4 percent of transactions cause over 50 percent of the rework. Four percent of Americans have over half the wealth.

Better still: The research into the diffusion of innovation shows that true transformational change begins with less than 5 percent of the workforce (the 4/50 rule). It also suggests that to accelerate results you will want to reduce the number of people involved.

> **Tip: To increase Lean Six Sigma results,
> reduce the number of people involved. Take the low road.**

Don't confuse activity with results! It doesn't matter how many people you've trained or how many teams you've started. That's just activity, not results. To accelerate Lean Six Sigma, narrow your focus.

The business world seems to be increasingly divided between the haves and the have-nots, the Lean Six Sigma snobs and the plebeian masses. The reigning wisdom seems to be that to succeed at improvement, you have to embark on a total cultural transformation.

Sadly, I haven't heard anyone talking about the benefits they have achieved from implementing such a transformation. There seems to be this illusion that if you embark on improvement, you'll magically be transported to a place of productivity and profitability. Nothing can be further from the truth. I've heard too many stories of massive investment with little return. One quality auditor expressed concern that if we aren't measuring the return on investment (ROI) of Lean Six Sigma, we're just fooling ourselves. After you pony up an estimated $250,000 (for training, salary, projects, and the like) to develop a Lean Six Sigma black belt, are you going to get at least $50,000 a project?

So why are all these big companies trying to do it the all-or-nothing way? Because you can't be criticized for aggressively doing everything possible to implement Lean Six Sigma.

There Has to Be a Better Way

There is a better way that produces better results with minimal risk: I call it the *crawl-walk-run strategy*. First, use the power of "diffusion" to implement Lean Six Sigma. *Start small*, with the first 4 percent of your business that produces over 50 percent of the waste and rework, then the next 4 percent, and so on until you reach a critical mass. Then Lean Six Sigma will sweep through the company, pulled forward by word of mouth. When I explain this crawl-walk-run approach to business owners, each one seems to awaken from his or her fog of despair and envision a path to Lean Six Sigma that is doable.

Set BHAGs

Conventional wisdom suggests that the goal is incremental improvement. However, if 4 percent of the business can produce over half the lost productivity and lost profit, why aren't you shooting for what author Jim Collins (*Built to Last* and *Good to Great*) calls a "Big Hairy Audacious Goal" (BHAG).

Set a BHAG to reduce delay, defects, or variation in one of your mission-critical systems by 50 percent in six months. Set a BHAG to reduce cycle time in a customer-critical process by 50 percent in the next six months. You'll be surprised how far such a goal will take you.

Use SWAT Teams

Instead of letting teams choose their focus, consider two-day leadership meetings to define and select key objectives. Instead of teams that meet indefinitely, consider using data to narrow your focus and having one-day root-cause "meetings" that bring together the right internal experts to focus on solving a critical business problem that affects customers and therefore profitability. These meetings focus on analyzing and verifying the root causes of problems and then identifying solutions.

There are *instant* solutions that can be implemented immediately by the meeting participants, and there are *managed* solutions that need some leadership and project management to ensure proper implementation. Instead of widespread training, only train teams that have a real problem to solve.

Right-Size Your Lean Six Sigma Team

The June 12, 2006, issue of *Fortune* magazine focuses on the secret of greatness—teamwork. It offers insights into teams past, such as Apple Computer's Macintosh team and teams of Marines in Iraq. It also argues that "most of what you've read about teamwork is bunk." While you can't just demand teamwork, there are some simple lessons:

- Size matters.
- Stars try to outshine each other.
- It's what you know *and* who you know.
- Location matters.
- Motivation matters.

The Marines' "recon" teams consist of six men. Jeff Bezos at Amazon has a "two-pizza" rule: If a team cannot be fed by two pizzas, it's too large. A professor at Harvard, J. Richard Hackman, bans student project teams larger than six. Hackman and Neil Vidmar found that the optimal size for a team is 4.6 people (think the Beatles plus their manager Brian Epstein). They also found that the minimum team size is three (two is a partnership). Another model for team size is the number of actors on *Gilligan's Island* or *Friends* (six to seven).

With three people, there are three communication paths. With four, there are six paths. With five, there are 10 paths, and so on. Too many paths results in delays and errors in communication that lead to delays and defects in the team's solutions.

Dream Teams Can Be Nightmares

Star players often try to outshine each other, leading to conflict, not collaboration. The relatively unknown cast of the movie, *My Big Fat Greek Wedding*, outperformed *Ocean's Twelve* with a cast of top actors. Sports dream teams sometimes can't play well together. Want to make some progress? Convene a team of knowledgeable but nonstar performers.

Leverage Your Centers of Influence

As Malcom Gladwell identified in his book, *The Tipping Point* (Little Brown, 2002), there are people in your company who are the true centers of influence. They may not have the top job, but they do have the ear of the people. They can make or break your success. There are two types of centers of influence: connectors and mavens. Everyone comes to the maven for their encyclopedic knowledge of the business or technology. The connector knows everyone and succeeds by connecting the right resources. It would be a good idea to engage your connectors and mavens in the improvement team.

It's Hard to Think Outside the Box
When You're Still in the Same Old Box

Lockheed had the "skunkworks." So did Ford's team Taurus and Motorola's team Razr. Sometimes you have to get out of your work environment to disengage the forces shaping your thinking. Get out of the building. Find a park bench or a hotel conference room or someplace that doesn't constantly remind you of the status quo.

Enhance Team Dreams

The best motivator may be impending doom or a fierce competitor. Then team members work together to serve the common good, as did Motorola's Razr team. Teams can bond to serve a stellar vision of the future, as did Apple's Macintosh or iPod teams. Whether you're defeating a foe or reaching for the stars, high-performance teams need something to move away from or toward—something that really matters to them and to the company. Otherwise, there's little motivation to survive or achieve.

Lean Six Sigma Is Easy; Teamwork Is Hard

For those of us who have been around Lean Six Sigma for a while, we know that the methods and tools are easy. It's the people and culture stuff that's hard. This is one of the main reasons that I recommend that people focus on the 4 percent of the business that's causing over 50 percent of the delay, defects, or deviation and engage only the employees involved in that 4 percent.

I also recommend that the teams be no larger than five to nine people. When focused on the 4 percent, a handful of people usually can solve any problem in half a day or less, whereas a wider focus and more people often ensure failure.

Teamwork is important to the success of the team, but as they say, it's "like getting rich or falling in love, you cannot simply will it to happen. Teamwork is a practice. Teamwork is an outcome." And teamwork leverages the individual skills of every team member. What can you do now to maximize your team's success?

WHAT'S WRONG WITH MOST LEAN SIX SIGMA TRAINING?

> Training driven change is built on two false assumptions: (1) individuals can change the system and (2) individuals can apply (successfully) what they've learned.
>
> —KEN MILLER

It's about classroom training, not experiential, on-the-job learning! To put this another way, a way that may be unpalatable for many of you, classrooms are, for the most part, a waste of time.

> You cannot learn anything in a classroom that is procedural in nature.
>
> —ROGER SCHANK

Think about the most useful things you've ever learned. Did you get them in a classroom or through actual practice in the real world? When I look back, it's a little bit of just-in-time learning with some expert coaching and a lot of practice.

In his book, *Lessons in Learning, e-Learning and Training* (John Wiley & Sons, 2005), Roger Schank examines the limitations of classroom training and the power of experiential learning. Here are some of his insights:

1. *Classroom LIBITI—Learn it because I thought it.* Schank said, "Consider Euclidean geometry. You have to agree that Euclidean proofs don't come up much in life. We tend to justify learning such things because we imagine that scholars have determined that the thoughts of great

thinkers ought to be learned. The reality is that if we were really concerned with [how this applies to your job] we would teach geometry in the context [of your job], not worrying about proofs so much as worrying about getting the measurements right." The same is true of Lean Six Sigma. You don't need to know how to calculate the statistical formulas for control charts, but you do need to know which chart to choose and how to read the result.

"As absurd as it is in school, LIBITI *is downright crazy in corporate training.*"

Principle 1: Just-in-time delivery makes information useful (don't tell people things that they cannot make use of immediately). **Lean principle:** Avoid overproduction.

Change happens in projects.

—KEN MILLER

Principle 2: Authentic activities motivate learners (don't tell people how to do something they will never have to actually do in real life).

Unfortunately, most Lean Six Sigma training isn't connected to the business. One Lean Six Sigma consultant admitted to me that he had three days of Design of Experiments (DOE) training in the healthcare black belt curriculum even though healthcare would rarely need such a skill. This is a form of *overproduction*—teaching things that people won't need.

2. *Things that are true when learning takes place:*
   - There is a goal that learning will help us to achieve.
   - The accomplishment of the goal is the reward.
   - After a skill is learned, it's practiced every day for the rest of your life.
   - There is continuous improvement.
   - The skill enables independence.
   - Rewards that accrue from future use are unknown at the time of learning.
   - Failure isn't a problem because failure occurs with nearly every attempt to learn.

- The process is not overly fun, but neither is it terribly painful or annoying.

    When I conduct just-in-time learning, I focus on applying everything I teach to healthcare and hospitals. Not some mythical pizza joint. You can learn the essential tools and methods in a day. My goal is to help you to develop your first improvement stories *during* the class. With improvement teams, I use 1 hour of just-in-time training to lay the groundwork for what team members will experience over the next day or two.

3. *How do learning designers do this?* Make sure that
    - Training is a group process.
    - Training is a problem-solving process.
    - Whatever is learned is merely a prelude to lifelong learning.
    - Make sure that independent use of the learning is in sight.

4. *How do we do this?*
    - Ask the experts what goes wrong in their companies.
    - Start thinking about training as just-in-time problem solving.
    - Start thinking about learning as doing, not memorization.

5. *Use stories in training.* The unconscious learns from stories. Stories should be about a particular attempt to slay a particular dragon.
    - Use real improvement stories.
    - Never "tell" without using a story.
    - Make sure that the tellers are authentic.
    - Tell stories just in time.
    - Relive the story, don't just tell it.

Simple truth: People learn by doing.

What's the real problem with most training? Too much training, not enough doing! Ninety percent of what you learn in a Lean Six Sigma class is lost if you don't apply what you've learned within 72 hours. This means that by Thursday of a five-day green belt class, you've forgotten what you learned on Monday. By the following Monday, when you go back to work, you've forgotten most of what you learned the previous week. Overproduction (e.g., training) produces unnecessary inventory (of methods and tools) that confuses the participant.

Why Most Corporate Trainers Fail to Teach by Doing

1. Real-life is too hard to replicate in a classroom.
2. It takes too long.
3. No experts are available for one-on-one help.
4. They want to teach general principles.
5. The subject matter doesn't seem "doing-oriented."
6. The training department has a list of learning objectives that can be learned *without* doing. Learning objectives tend to trivialize complex issues by making them into sound bites that can be told and then tested.
7. They don't know how to do it.

Doing-based learning involves

- Practice
- Feedback
- Reflection

People learn best when they are pursuing goals that they really care about and when what they learn directly helps them attain their goals. The best means of learning has always been experience.

—ROGER SCHANK

People learn best when they

1. Experience a situation
2. Must decide how best to deal with the issues that arise in that situation
3. Are coached by experts

This is the essence of how I teach Lean Six Sigma, through stories, examples, and the essentials, not every little detail; you can too. Once you understand what I call the "spine," or the essence of knowledge, you can add to it forever. You can look up the more exotic requirements of Lean Six Sigma when you hit a situation that requires them.

Can you learn *everything* about Lean Six Sigma in a day or less? No. Can you learn everything you need to know to make dramatic progress from three to five sigma? Yes, you can. Can you really afford to send your employees to weeks and weeks of training? Maybe if you have deep pockets, but not if you want to get started and make progress. Sign up for our free money belt training at www.qimacros.com/moneybelt.html.

ARE YOU A LEAN SIX SIGMA SALMON?

In 2003, the Benchmarking Exchange conducted a survey of Lean Six Sigma companies. The first question the surveyors asked was, "Within the past 24 months, what business processes have you or your company targeted for improvement?" Top answer? Customer service and help desk—the mouth of the river of defects and delay, not the source!

Most companies make the same mistake. The pain they feel is in their customer-service and help-desk areas. Too many calls. One wireless company I worked with received 300,000 calls a month from only 600,000 subscribers. Ouch!

But the root cause is rarely in the customer-service center; it's somewhere *upstream*: incorrect orders, order fulfillment, service delivery, billing, and so on. The customer-service center is a major piece of your company's "fix-it factory." It's also a source of excellent data for improvement projects.

In my small SPC software business, I consider every customer "service" call to represent a defect. Let's face it, if every customer-service call costs $8 to $12, how many calls do you want to take? *None*, right?

So I ruthlessly try to find ways to make the installation and operation of the software painless and effortless. I try to put all the answers a customer will ever need on my Web site so that they can serve themselves when I'm not available. I try to mistake-proof everything in every interaction.

The wireless company I mentioned had set up its entire business to systematically herd customers into the customer-service center. The company assumed that its customers knew nothing about cell phones and would need to call the company to learn how to use them. Every piece of documentation directed customers to call. The monthly bill with roaming charges and extra minutes drove people to call. It was a nightmare.

My business, on the other hand, is more like the Maytag repairman commercials. I *never* want a customer to have to call. What caused most of my calls? When I looked at the data, it was ordering, not the software. So I streamlined and mistake-proofed the ordering process.

Of course, it's possible to go overboard. Look at Microsoft. I can't figure out where to call to get help when I need it. There's an enormous knowledge base at http://support.microsoft.com, but I can't always find what I need there. It seems like every day I get a new message that there's some new Windows update waiting to be installed. (This makes me think that Microsoft's software is awfully buggy. Does it really need daily updates?)

Get the Order Right!

I must have bad service karma or the quality crisis is growing. In the last couple of weeks, I've been to Chili's and Wendy's and failed to get what I ordered several times.

Chili's. My wife and I went to Chili's and ordered the chicken and shrimp fajita for dinner. We got a plate of chicken and shrimp, but it wasn't a fajita. The manager asked if we wanted to wait for the kitchen to cook up the fajita, but we were going to a movie, so I said no. The manager offered to comp our meal, but later our server said that she'd offered to give us a free dessert. Not only couldn't the company get the order right, it couldn't get the compensation right either.

A couple of weeks later, after avoiding Chili's, we decided to grab some takeout. I again ordered chicken and shrimp fajitas, but when we got home, we only had shrimp, no chicken! Wrong order again!

It costs time and money to correct an inaccurate order. It costs even more money—intangible money—because customers avoid buying from you.

Wendy's. I've driven through Wendy's a couple of times and ordered a number 1, which is a hamburger with a side dish and a drink. Wendy's offers fries, salad, chili, or a baked potato. Each time I ordered a salad with ranch dressing. Each time, when I got to the window, I received a hamburger, fries, and a drink. Most of the time I have to remember to check my order to make sure I get what I ordered. But why should I do Wendy's quality control? It irritates me.

When I remember, the server gives me the salad, but I end up keeping the fries because the company can't take them back. Even a few fries cost money. And I hold up the drive-thru line waiting for my salad.

Wrong order, wrong product, angry customer. Is the company so busy trying to remember to "supersize me" that it can't actually hear the order?

Lil Ricci's Pizza. Near our house is a New York–style pizzeria. The servers at Lil Ricci's, no matter what kind of curveball I throw at them—hold the onions, extra mushrooms, or whatever—always get it right, and they do it by memory. Not only is the pizza great, but the service is excellent. And I tend to overtip because they get it right.

And Restaurants Aren't the Only Culprits. Yesterday I spoke with a friend who had a new house built. The builder put in the wrong kind of hardwood for the floors. My friend made them take it out and put in what she ordered. Ouch! There went the profit on that house.

I also worked with my mail-order house yesterday. It has too many tales of woe about not getting an accurate layout of how to print addresses for its clients. Then the company has to reprint on labels and manually paste the new address over the old address. And the company sometimes duplicates mailings or uses the wrong address files. The company has checkers checking the checkers, but errors still slip through. And I know that my printer routinely writes off $100,000 a month in adjustments because he didn't get the print job right.

We Don't Always Get It Right. Even in my business, we don't always get it right. But we do stop to analyze why we didn't get it right and search for ways to mistake-proof the order process.

Here's My Point

If you don't get the order right (e.g., pharmacy, lab, radiology, operation, etc.), nothing else will matter. There's rework, waste, and extra cost on your end. There's rework, waste, and extra cost on your customer's end. There are bad feelings that discourage future transactions.

And it's not your people's fault. Your systems and procedures let them make mistakes. Change your processes, procedures, and information systems to prevent incorrect orders. Make it impossible to enter an order incorrectly!

Wouldn't it be easier to spend a couple of hours figuring out how to prevent order errors than dealing with the seemingly endless rework, waste, and cost they cause? Wouldn't it be easier to spend a couple of extra seconds *listening to your patient* and getting the order right?

It's up to you, but haven't you waited long enough to start shifting how you do business so that you delight patients and earn more trust and more business?

Look Upstream

The customer-service center is a great place to gather data about the customers' problems, difficulties, and issues. It is a terrible place to try to *solve* those problems and issues. Analyze the data from your customer-service center, and then initiate root-cause teams in the appropriate departments to *solve the upstream problems* that are drowning your customer-service help desk.

Don't be a salmon! Start at the source. Clean up the sewage at the headwaters of your business. Keep analyzing why customers call, but use that information to fix operations, not the call center. Sure, call centers need improvement too, but if customers don't need to call, do you really need a call center? Maybe, but does it need to be as big as it is?

- Mistake-proof your operations, products, and services.
- Simplify your product or service.
- Make more things self-service.

Spring Forward—Fall Back

This mantra of Daylight Savings Time that I learned as a kid seems to hold true for Lean Six Sigma as well. Back in February, *Quality Digest*'s survey found that most large companies were springing forward with Lean Six Sigma and then falling back two to three years later when time, money, lack of ROI, or a change of CEO cast a shadow over the promise of Lean Six Sigma.

I spent some time with a large power company that had actively pursued quality in the 1980s under the leadership of one CEO, only to reverse course during the 1990s under another CEO. Sadly, most of the skilled quality personnel left during that period. And now, the company is returning to quality under the Lean Six Sigma umbrella and is using a version of my crawl-walk-run approach to do it.

Rather than invest in massive training programs, the quality staff is quietly finding operational vice presidents who want to cut costs and boost profits. Then they use the methods and tools of quality to make those improvements and convince the vice presidents that Lean Six Sigma can help them (1) get ahead personally and (2) move the company forward as well. This develops buy-in to create additional projects and weave quality into the fabric of the department.

Has your company jumped forward? Are you falling back? Are the returns less than expected? Is your leadership changing?

"Make It Sticky!"

This is the title of Chapter 15 of Jack Welch's book, *Winning: The Answers* (HarperCollins, 2006). He says that Lean Six Sigma "can be one of business's most dreary topics." But he also says

- "I am a huge fan of Lean Six Sigma."
- "Nothing compares to the effectiveness of Lean Six Sigma when it comes to improving operational efficiency."
- "The biggest but most unheralded benefit of Lean Six Sigma is its capacity to develop a cadre of great leaders. It builds critical thinking and discipline."
- "Lean Six Sigma is one of the great management innovations of the last quarter century and an extremely powerful way to boost a company's competitiveness."
- "You can't afford not to understand it, let alone not practice it."

Yeow! "Yet, for many people, the concept of Lean Six Sigma feels like a trip to a dentist."

Welch offers some insights about how to coat Lean Six Sigma with Teflon (how *not* to do it):

- "Hire statisticians to preach the gospel."
- "Use complex PowerPoint slides that only an MIT professor would love."
- "Present Lean Six Sigma as a cure-all for every nook and cranny."
- Welch's advice: "Don't drink the Kool-Aid!"

As I've argued since its inception, at its heart, Lean Six Sigma is simple. You only need to know a few key methods and tools to make huge progress in most companies. Eventually, you'll need to learn more robust methods, but not right out of the gate.

Creating Stickiness with SUCCES. Dan and Chip Heath wrote an excellent book on making ideas stick, *Made to Stick* (Random House, 2007). They found six key principles of sticky ideas (acronym SUCCES):

- *Simplicity.* Ideas must be stated simply and also be profound.
- *Unexpected.* Ideas must be surprising.
- *Concrete.* Ideas must be stated concretely in ways that you can see, hear, feel, smell, taste, or touch. Sadly, Lean Six Sigma is full of jargon.
- *Credible.* Ideas must be believable.
- *Emotional.* Ideas that stick invoke a feeling or emotion that acts like Velcro in the mind.
- *Stories.* Stories, especially mysteries that reveal the solution at the end, are especially powerful for helping ideas stick in the mind.

This is why I like to tell Lean Six Sigma improvement stories as a way to train team members. A simple, concrete improvement story is the best way to convey the Lean Six Sigma improvement methods and tools. It engages the right brain, not just the left. And the methods and tools leave a lasting impression that sticks.

Think about the things you remember from childhood. Most are jingles or stories. I can't remember a single thing about differential equations, but I can remember an episode of *The Lone Ranger*. That's the power of simple, concrete stories. They stick.

The Elevator Speech. Most sales and marketing books recommend that you develop an "elevator speech" (one that can be given in a 30-second elevator ride) about your business. Jack's elevator speech about Lean Six Sigma is "Lean Six Sigma is a quality program that improves your customers' experience, lowers your costs, and builds better leaders." Or more simply, "the elimination of unpleasant surprises and broken promises." Here's mine: "I work with companies that want to plug the leaks in their cash flow." Develop one of your own: Lean Six Sigma lowers costs while boosting profits and productivity.

Get "Sticky."   Variation in defects, delay, and cost make your business unpredictable, and customers hate unpredictability. Americans love fast-food chains for their predictable menus, quality, and speed. This is part of what Welch calls "stickiness"—creating products and services that make the customer come back time after time.

Fat Cats Don't Hunt

Most hospitals are making a profit at around 3 to 4 sigma. They have no idea how much profit they could make if they started moving toward 5 sigma. They have no idea how much better their patient outcomes would improve. And you don't need a flock of black belts and green belts to get going, just "money belts."

Given the choice between developing excuses about why they can't improve or applying the basics of Lean Six Sigma to measure and improve defects, delay, and cost, though, most people get busy on the excuses. You can make huge progress in 6 to 12 months. Wait a year, and you risk letting your competition get a head start on creating a "sticky" product or service. As the U.S. automotive industry discovered, it's hard to catch up once you're behind.

Here's my point: Lean Six Sigma isn't like a trip to the dentist; it's like a trip to the bank to deposit a wad of cash. Use it.

RISK-FREE WAY TO IMPLEMENT LEAN SIX SIGMA

Over half of all TQM implementations failed. In the language of Lean Six Sigma, that's 1 sigma, a pathetic track record. And if you study how most companies are implementing Lean Six Sigma, you'll find the same old formula that ruined TQM:

1. Get top leadership to commit to widespread implementation.
2. Train internal trainers (black belts) to minimize the costs of training everyone.
3. Internal trainers train team leaders (green belts).
4. Start a bunch of teams.
5. Hope for the best.

Everyone points to General Electric as a leader in Lean Six Sigma, but if you look more closely, you'll see that Jack Welch had already created a company that managed and even embraced change. Implementing Lean Six Sigma, therefore, wasn't as hard as it might be in other organizations.

Many people I talk to in various industries say that they tried TQM and it left a bad taste in their mouths. So Lean Six Sigma not only has to overcome resistance to change but also the bad taste left by failed TQM implementations.

So how do you implement Lean Six Sigma in a way that's risk-free? By using the proven power of diffusion [*Diffusion of Innovations* (Free Press, 1995), by Everett Rogers]. Over 50 years of research into how changes take root and grow in corporations and cultures suggests a much safer route to successful implementation of Lean Six Sigma or any change.

The employee body can make three choices about Lean Six Sigma or any change: adopt, adapt, or reject. In a world of too much to do and too little time, rejection is often the first impulse. People rarely adopt methods completely, so there must be room for adaptation to fit the corporate environment. There are five factors that affect the speed and success of Lean Six Sigma adoption:

1. *Trialability.* How easy is it to test drive the change?
2. *Simplicity.* How difficult is it to understand?
3. *Relative benefit.* What does it offer over and above what I'm already doing?
4. *Compatibility.* How well does it match our environment?
5. *Observability.* How easy is it for leaders and opinion makers to see the benefit?

You also can speed up adoption by letting the employees decide for themselves to adopt Lean Six Sigma rather than having the CEO decide for them (although this is how we keep preaching success—get the CEO to commit to widespread change).

Thus, to maximize your chance of success and minimize your initial investment:

1. *Start small.* Forget the 80/20 rule. Less than 4 percent of any business creates over half the waste and rework. So you don't have to involve more than 4 percent of your employees or spend a lot of money on widespread training to get results. Get an external Lean Six Sigma "money belt" to help you find and create solutions using the tools and methods of Lean Six Sigma. Your employees will learn through experience, which is far more valuable than classroom training.

2. *Set BHAGs.* Go for 50 percent reduction in delay, defects, or costs. When you're just starting out, big reductions are often easier to get than you might think, so why not go for them? This also telegraphs the message to your teams that this isn't continuous improvement.

3. *Fly under the radar.* Most companies broadcast their Lean Six Sigma initiative, and employees think, "Here comes another one." This usually stirs up the laggards and skeptics—what I call the "corporate immune system." You are much better off to make initial teams successful and let the word of mouth spread through informal networks because this is the fastest way for cultures to adopt change.

4. *Create initial success.* In 1980, the company I worked for brought in a trial of 20 Time-Sharing Option (TSO) terminals (to replace the punched cards IT used). The company selected a small group of programmers to use the terminals. The buzz from this one group caused TSO to be immediately accepted and integrated into the workforce. Do the same thing for Lean Six Sigma. Only start teams that can succeed. Make a small group of early adopters successful. Then another, then another.

   When the pioneers (early adopters) become successful, they will tell their friends. The pioneers will convince the early settlers, who eventually will convince the late settlers. No one will ever convince the laggards and skeptics; they have to convince themselves.

5. *Fight the urge to widen your focus.* Remember the dark side of the 4/50 rule: 50 percent of your effort will produce only 4 percent of the benefit.

6. *Simplify.* Using simple tools such as control charts, Pareto charts, and fishbone diagrams, you can easily move from 3 to 5 sigma (30,000 parts per million to only 300) in 18 to 24 months. There are lots of more sophisticated tools, such as QFD and DOE in Lean Six Sigma, but you won't be ready to design for Lean Six Sigma until you simplify and streamline your existing processes and lay the groundwork for it.

7. *Review and refocus.* Once you solve the initial 4 percent of your core problems, start on the next 4 percent and then the next. Diffusion research has shown that somewhere between 16 and 25 percent involvement will create a "critical mass" (Figure 12.2) that causes the change to sweep through the culture.

The Good News About Productivity and Profitability

When you focus on the 4 percent that creates over half the waste and rework, your initial teams get big benefits: 50 percent reduc-

FIGURE 12.2

Adoption of Lean Six Sigma.

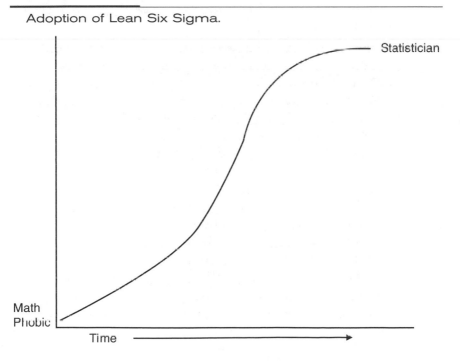

tion in defects, waste, and rework and $50,000 per project improvement in the bottom line. By the time you've worked your way through the first 16 to 20 percent of your problems, you will get 80 percent (the 80/20 rule) of the benefits of Lean Six Sigma. And you'll have minimized your costs of implementation. Now you can grow skilled internal black belts from your initial improvement team members. And you can begin to think about Design For Lean Six Sigma (DFLSS) to design your processes to deliver Lean Six Sigma quality.

Lean Six Sigma payoffs are huge, but you may want to consider using the power of diffusion to ensure that the methods and tools take root in your business and flourish. But it's up to you. You can choose the conventional wisdom that gives you only a 50:50 chance at success or choose the power of diffusion that increases your odds substantially. Haven't you waited long enough to start making breakthrough improvements in performance and profitability a permanent part of your business?

Prerequisites for Lean Six Sigma.   To succeed at any Lean Six Sigma project, you need a few key things:

1. *A project worth doing.* This means that it should be worth at least $50,000 in savings. If it's not worth doing, find a better project.
2. *A project concerned with operational problems that you can control directly.* You can't directly affect customer perceptions of your business, but you can improve your speed and quality. You can't make customers talk about your product or service, but you can improve it so much that they can't help but brag about it.
3. *Available data about the project.* This means that you need measurements of the problem over time. And you need the underlying measures of various contributors to the problem to be able to laser-focus your analysis. If you don't have the data, you will have to start collecting it. But this takes time. Isn't there a different $50,000 project that has all the data you need to get started right now?
4. *An operational manager or leader who wants to solve the problem.* Without sponsorship, most projects will fail because you won't get the time and resources you need to succeed.

5. *An experienced Lean Six Sigma guide.* This can help you to laser-focus the effort, find the root causes, and implement solutions in a matter of days, not months.

If you don't have one of these prerequisites, spend the time to change your focus or get what you need to move forward. Otherwise, you're just wasting your time, and you're doomed to failure.

But when you meet these prerequisites, your chances of success soar. When you have a worthwhile project, the data to support it, committed leaders, and an experienced guide, you can get results in days or weeks, not months or years.

Defending Your Data

Another Six Sigma tar pit involves data. Everybody likes to feel good about their job and themselves, and nobody likes to feel bad. This is one of the major challenges of quality improvement. Most people would prefer to focus on what's going well rather than on fixing what isn't quite working.

Sadly, when it comes to using facts and figures to improve the business, most people get busy trying to cast a shroud of suspicion over the data to discredit them and avoid doing anything. Almost daily we get calls from QI Macro users who are trying to prepare for the onslaught of criticism they're sure they'll face when presenting their data, charts, and graphs to a "higher power." Nurses tremble when facing doctors. Employees worry when presenting to the boss. Most employees aren't statisticians, just people trying to do a good job for a customer, but they worry that someone will challenge their lack of understanding of math and statistics. Here are some of the common issues I hear about. Let me know about yours.

The Data Aren't Perfect. And it can happen to anyone. In March 2004, a report by the U.S. Centers for Disease Control and Prevention (CDC) concluded that poor diet and lack of exercise were responsible for 400,000 deaths in 2000, up 33 percent from 10 years earlier. In November 2004, the *Wall Street Journal* reported that the number may have been overstated by 80,000 because of mathematical errors such as including total deaths from the wrong year.

The CDC acknowledged that there may have been statistical miscalculations in the report. The agency plans to submit a correction to the *Journal of the American Medical Association*, which published the original study. Even with the corrections, obesity remains the second leading cause of preventable death.

All data are imperfect. Get over it. You can make a lot of progress using imperfect data.

The Data Aren't Valid. This is the easiest way to throw the hounds off the scent.

> *Ask:* Do you have better data? Show me. (Most of the time, they won't.)
> *Say:* Until you bring us better data, we'll have to move forward with what we have.

Why Don't We Measure Our Successes Rather than Our Failures? People want to feel good about what's going right, but improvement is about reducing mistakes, defects, and errors. So you have to focus on the failures. Prevent the failures, and success will improve automatically.

I Don't Like the Answer. When you start using Pareto charts, control charts, and other documents that actually reveal the extent of a problem, people won't feel good about it. The fastest way out of feeling bad is to discredit the data or the person who brought them up. I've even heard managers say the phrase *wrong answer*.

When people use a control chart, they find that the process is unstable and needs improvement. "You must be using the wrong chart," they proclaim. Many of these naysayers know how to sound confident and are competent enough to make the presenter doubt his or her data. Don't buy it.

> *Ask:* Show me what's wrong with it. What chart would you recommend? Let's draw it now! (And you can using the QI Macros.)

I Don't Get the Same Answer—The Formulas Aren't Right. Some bosses want their people to verify the QI Macros by creating their own formulas and spreadsheets, and then they wonder why their 15-minute effort doesn't correspond with the software I've been developing for a decade.

Just because the QI Macros aren't the most expensive piece of SPC software in the world, some people think that they're cheap (i.e., poorly constructed, inaccurate). "Wrong answer!" The formulas in the QI Macros have been endlessly tested and come from the most up-to-date statistical references (such as Juran's *Quality Control Handbook*) and standards groups (such as the Automotive Industry Action Group [AIAG]).

More often than not, the user just misinterpreted the formula. I had one customer fiddling with the formulas for Cp and Cpk. He got the formulas off a Web site (and they were correct), but he missed the little bar over the *R* for range that means the *average of the ranges*, so he used the maximum minus the minimum to get a range and then chose the wrong value for n to calculate sigma estimator.

> *Ask:* What formulas are you using? What reference book are you using?
>
> *Say:* The formulas are fine. If you want to know more about the formulas, buy a copy of a good SPC book. Meanwhile, what are the data telling us?

Why Are There So Many Control Charts?   Why isn't there just one? Why don't you just use standard deviation? Aren't the upper centerline and lower centerline supposed to be ±3 standard deviations away from the mean? This is another example of people not understanding some basic stuff.

> **The answer:** You could use standard deviation if you have all the data and they're normally distributed, but when you use samples or have different kinds of distributions (e.g., defects), the formulas vary to account for the differences.

My Statistics Book Doesn't Match Your Statistics Book.   One customer asked why Breyfogle's GageR&R example came up with different results from the QI Macros. On investigation, Breyfogle clearly got his information from the *AIAG Measurement Systems Analysis* (Second Edition, 1995), whereas I'm using the third edition. Donald Wheeler has a slightly different control chart constant (3.268) for one kind of chart than every other SPC reference book (3.267), but let's face it, 0.001 isn't going to change the control limits by more than a hair.

Another thing I've noticed is that every author has to change the symbols or the layout or something to avoid looking like they copied the stuff from another source. The same customer asked me why the formulas in one reference book weren't in the macros.

Would I consider adding them. On further investigation, I found that the formulas are there in a format different from mine. No wonder it gets so confusing.

One Bad Apple.   Many customers have created a histogram and then wondered why they have one big bar on the left side and a small bar way out on the right (or vice versa). A lot of data get entered manually. I usually find that one data point is entered with the decimal point in the wrong place. For example, you may see data in the form of 0.01, 0.03, 3.0, and 0.02.

Ask: Have you checked your data?

Dummy Data.   There's an old saying in information technologies: "Garbage in, garbage out (GIGO)." Several customers have put "dummy data" into tools and then been caught off guard because the template tells them their system needs improvement. Dummy data can lead to dummy results.

Ask: Where did these data come from?

Preprocessing the Raw Data.   Several users have sorted the data before drawing a histogram, which affects how the ranges are calculated and really screws up the Cp and Cpk calculations. Other customers turn their raw data into ratios or averages but then try to use the ratio in a chart that needs the raw data. Many healthcare clients take ratios such as falls per 1,000 patient-days but then try to use the ratio in a p chart that needs the raw falls and patient-days. Another person tried to use two averages to do a statistical t test.

Ask: Have you done anything to the raw data?

Focus on the Goal, Not the Methods or Tools or Data.   You can make a lot of progress with imperfect data. Improvement projects only need "good enough" data, not perfect data. Stop using your data as a crutch to avoid fixing important problems. Stop using your charts as an excuse to argue about statistics and tools. Instead, ask yourself, "What can we learn from this chart or graph? What are the data telling us? Is there a problem worth solving? Where should we focus our improvement effort?"

Want to feel good again? Improve some mission-critical process by making it far better, faster, and cheaper than ever before. This will make you feel good. Stop haggling about data and formulas. Start making some progress on real business goals.

CAN LEAN SIX SIGMA KILL YOUR HOSPITAL?

Yes, it can. Peter Keen, in his book, *The Process Edge: Creating Value Where It Counts* (Harvard Business School Press, 1997), uses case studies to describe what he calls the *process paradox*. The process paradox: "Businesses can decline and even fail at the same time that process reform is dramatically improving efficiency by saving the company time and money and improving product quality and customer service."

> Continuous improvement is the right idea if you are the world leader in everything you do. It is a terrible idea if you are lagging and disastrous if you are far behind. We need rapid, quantum-leap improvement.
>
> —PAUL O'NEILL, CHAIRMAN OF ALCOA

Wrong Implementation

Most companies are trying wall-to-wall, floor-to-ceiling implementations of Lean Six Sigma. Sadly, this means that 80 percent of the people are engaged in trying to get less than 20 percent of the benefit. Wall-to-wall implementations can siphon valuable resources away from satisfying customers, creating new products, and exploring new markets.

Wrong Process

Invariably, most Lean Six Sigma teams want to start with a pilot project that's not too risky. Unfortunately, they end up majoring in minor things. They don't get the results required to make a case for Lean Six Sigma.

Wrong Team

Invariably, leaders try to form a Lean Six Sigma team before they've analyzed the data to figure out who ought to be on the team. Consequently, the team struggles because team members are not the right people to solve the problem once it's been stratified to an actionable level.

> **Here's my point: Lean Six Sigma can kill your business just as easily as liberate it.**

Ruthless Prioritization (a.k.a. Laser Focus). Keen suggests that every business tries to boil the ocean or solve world hunger rather than narrowing their attention to a few customer- and profit-critical value-adding processes where they *can* make breakthrough improvements. Use the 4/50 rule to narrow your attention to the 4 percent of your business that causes over half the lost profit. Tackle the big hairy audacious problems in your business first.

ASSETS AND LIABILITIES

Keen suggests that every process is either an asset or a liability. It either adds value or it doesn't. He also suggests that there are five types of processes:

1. *Identity.* Processes that define the company to patients, employees, and investors.
2. *Priority.* Processes that are critically important to business performance.
3. *Background.* Processes that provide support to other processes.
4. *Mandated.* Required by law (e.g., taxes).
5. *Folklore.* Legacy processes that have no value.

Table 12.1 shows Keen's matrix for evaluating your existing business processes. How many fall into the liability and folklore cells?

TABLE 12.1

Assets and Liabilities

Process	Asset	Liability
Identity		
Priority		
Background		Repair and rework
Mandated		
Folklore		

Far too many departments and individuals think that fixing mistakes is an asset to the business. It's a liability because it eats profit and reduces growth. Isn't it time to narrow your focus to some mission-critical priority processes?

- Use data to narrow your focus to the 4/50.
- Stop training everyone. Train 4/50 team members just in time, right before they embark on the problem-solving process.
- Aim for breakthrough (50 percent reduction in defects or delay).
- Stop majoring in minor things.

Here's what I've observed: *Every company needs two key mindsets: (1) innovation and (2) improvement.* Lean Six Sigma is the best method around for making improvements when you have linear cause-effects. (Other methods are required when you start to investigate circular or systemic cause-effects.)

Breakthrough improvements in speed, quality, and cost often lead to streamlined and simplified products and processes that lead to innovative insights. The innovations need continuous improvements to survive and thrive. And the cycle of innovate-improve starts all over again.

The good news is that the *Lean Six Sigma mindset can benefit any company,* large or small, service or manufacturing, profit or nonprofit. The bad news is that you will need to keep reinforcing it forever so that you keep springing forward and rarely fall back. And the crawl-walk-run approach described in *Lean Six Sigma Simplified* is the best method I have found to take baby steps with Lean Six Sigma that produce giant leaps in performance.

This is what the science of complexity and the system thinkers call a *vicious reinforcing loop.* Tiny causes have big effects that become self-reinforcing. They lead to more small steps that deliver big results and so on. As you do this, you will systematically weave Lean Six Sigma into the fabric of the business, making it hard to rip out with each successive change of leadership or change in market conditions.

Lean Six Sigma isn't a panacea—a cure for all things—but it is extremely good at what it does well. Get the Lean Six Sigma mindset inside your company so that you can continue to spring forward and stop falling back.

Clayton Christensen, author of *The Innovator's Dilemma* (Harvard Business School Press, 2000), found that "management practices that allow companies to be leaders in mainstream markets are the same practices that cause them to miss the opportunities offered by 'disruptive' technologies. In other words, well-managed companies fail *because* they are well managed." And he offers many case studies from disk drive manufacturers and other industries that support his findings. What are the hallmarks of good management that cause companies to fail?

1. *Listening to customers.* Your current customers want more of the same from you. The emerging market doesn't know what it wants in the next big thing. It just wants simpler, cheaper, and more convenient products and services. Think iPod versus Walkman.

2. *Seeking higher margins and larger markets.* Not smaller emergent ones.

3. *Relying on market analysis to find new markets.* Markets that don't exist can't be analyzed; they can only be explored through trial and error.

Recently, I was asked to consult with a company that had been acquired by a larger company that was obsessed with Lean Six Sigma. The acquired company started training black belts and green belts and project teams. It required every employee to have his or her own Lean Six Sigma project. Every project started asking everyone else for data in all kinds of formats and layouts and selection criteria. The company literally became paralyzed doing Lean Six Sigma and forgot to take care of customers and continue its efforts at innovation, which is what the company was known for. Smaller companies were eroding its market share with simpler, more convenient, and cost-effective tools.

As Joel Barker said in his book, *Future Edge* (William Morrow, 1992), you "manage" within paradigms and you "lead" between paradigms. Lean Six Sigma is a great methodology and toolkit for managing and improving your product and service. Companies that do this continue to succeed as long as the underlying technology doesn't change dramatically. But when it does change, your hospital most likely will be incapable of recognizing and taking advantage of it.

Why? Because it means entering smaller markets with products that generate smaller margins until they become mainstream. IBM, for example, ignored minicomputers. DEC, which succeeded in minicomputers, ignored microcomputers. Apple computer started the microcomputer market, but IBM jump-started personal computers by creating a "skunkworks" to develop the first prototype. And none of these companies succeeded at developing the Palm Pilot, but the Palm failed to anticipate the iPhone. And so it goes.

Intel may be the only company that consistently rides the curve of new technology by using the strategies outlined in Christensen's book:

1. *Set up a separate organization.* It is small enough to get excited by small gains with customers who want the new technology.
2. *Plan for failure.* Make small forays and tailor the product or service as you learn what your emerging customers want. How many companies start down one path only to discover the big market is a derivative of the original idea?
3. *Don't count on breakthroughs.* Many next-generation technology markets emerge from recombinations of existing technologies. Smaller disk drives use the same technology as larger disk drives. So why didn't the large disk drive manufacturers spot the need for smaller drives in PCs, and why didn't PC disk drive manufacturers spot the need for still smaller drives for laptops?

Is Lean Six Sigma making you too complacent? Are you ignoring the tug of the emerging markets in your industry? Don't let Lean Six Sigma kill your company. Balance your efforts to improve the existing business and innovate for emerging markets. Stop the insanity. It's not either/or—improve *or* innovate—it's both—improve *and* innovate. Otherwise, your future is in jeopardy.

INNOVATION RULES

Innovation is clearly a success strategy for businesses in the information economy. Once thought to be the domain of only the creative and gifted, there appear to be some simple rules that encourage innovation.

Google Rules

Marissa Mayer is Google's innovation czar. In the June 2006 issue of *Business Week's IN-novation* quarterly, she lists her nine notions of innovation:

1. Ideas come from everywhere and everyone. Encourage them.
2. Share everything about innovation projects. Give everyone a chance to add to or comment on the process.

3. If you're brilliant, we're hiring. If your company thrives on innovation, you can't afford to pass up talent.

4. Give employees a license to pursue their dreams. Employees get one free day a week to work on whatever they want to work on. Half of new Google products come from this time.

5. Prototypes versus perfection. Launch early, test small, get feedback, and improve until you converge on the best product.

6. Don't BS; use data. Just because someone likes an idea doesn't mean that it's any good. As Motorola says, "In God we trust. All others must bring data."

7. Creativity loves restraint. Set boundaries, rules, and deadlines.

8. Worry about users and usage, not money. If you provide something simple to use and easy to love (see Google's home page or my QI Macros SPC Software for Excel), the money will follow.

9. Don't kill projects—morph them. Just like 3M's failed glue that made Post-it Notes possible, there's always a kernel of greatness in a failed project.

Types of Innovation

In *The Innovator's Dilemma*, Clayton Christensen identifies two types of innovation: sustaining innovation and disruptive innovation. Sustaining innovations such as Digital Subscriber Line (DSL) enables phone companies to carry more data over the same lines that carry phone service. Cell phones, however, are a disruptive innovation. Wires cease to be important when you can go wireless. Digital cameras make film cameras obsolete.

Fast Innovation

In Michael George's book, *Fast Innovation*, he suggests that every innovation effort has three imperatives:

- *Differentiation*—delivering a product or service that will touch the *heart of the customer*
- *Speed to market*—to gain first-mover advantage
- *Disruptive innovation*—to make your competitors obsolete

Rapid Prototyping

If a picture is worth a thousand words, we've found
that a good prototype is worth a thousand pictures.

—TOM KELLY OF IDEO

Speed to market and touching the heart of the customer rely on rapid prototyping of the product or service and testing it with customers in small pilot projects because people are better at reacting to prototypes than they are at coming up with ideas on their own.

Example. When I develop dashboards of performance measures for companies, I iterate several times to converge on the ideal layout for the company's measurement data. Then I *reuse* the templates in the QI Macros to create all their graphs.

RELIGION OF REUSE

Speed to market also depends on what George calls the "religion of reuse." Toyota reuses 60 to 80 percent of the designs and parts in new models of cars. This makes it possible to bring new models to market in half the time of competitors. You can too. This kind of information led George to formulate the 80/80/80 rule: 80 percent reuse will cut lead times by 80 percent at 80 percent productivity of the innovators, which results in shorter lead times (50 to 80 percent).

Higher productivity occurs because you can use smaller teams of highly focused individuals. Reuse can cover not just parts but documents and ideas as well. Keep a lookout for cool ideas. When Taiichi Ohno saw how American supermarkets stocked their shelves, he immediately saw a way to simplify and streamline inventory in Toyota manufacturing plants.

How does the religion of reuse apply to hospitals and healthcare? Simple: Adopt or adapt what other hospitals have already done to accelerate patient flow, reduce errors, and improve outcomes. Healthcare leaders such as Virginia Mason, Cleveland Clinic, and Mayo Clinic are leading improvement initiatives. What can you reuse from their learnings?

In the April 2010 *Harvard Business Review*, Oded Shenkar says, "Imitation is more valuable than innovation. Imitation is not mindless repetition; it's an intelligent search for cause and effect."

Scientists now view imitation as "a complex and demanding process that requires high intelligence and advanced cognitive capabilities." Ninety-eight percent of the value of any improvement or innovation is captured by the copycats.

> **Hint: Don't reinvent the wheel.**
> **Copy what others have already done.**

Simplify for Speed

Brooks' law says that adding people to a late project will only make it later because the communication costs go up exponentially. George says that to accelerate the innovation process, *reduce the number of projects* because you'll free up your critical innovation resources to focus their time on the key projects. One company that did so increased new products by 40 percent and reduced time to market by 40 percent.

Innovation isn't about cloning existing products and hanging a new name on them. Between 1996 and 1999, Procter and Gamble (P&G) reduced the number of "me too" product stock-keeping units (SKUs) by 20 percent, saving $2 per case, or $3 billion annually. The company cut the number of Head & Shoulders Shampoo SKUs by 50 percent, but sales per item doubled.

Measure Your Innovation Rate

As Marissa Mayer suggests, establish measures of innovation:

- Lead time for new products or services
- New products per year
- Revenue from new products per year
- Percent of product from reused components

Innovation is a mindset. It can be learned. There are some simple ideas such as prototyping and the religion of reuse that can be learned and applied immediately. What are you waiting for? Go out there and create the next big thing.

CONFLICTING GOALS

One of the biggest challenges with Lean Six Sigma is to align goals across the business. Take purchasing, for example. The goal of the purchasing department may be to get the best deal, but in so doing, the department may cost the company many more dollars than it saves.

One of my customers called and asked if she could still get the discount on 50 copies of the QI Macros if she bought them 25 at a time. Curious, I explored why she didn't want to buy all 50 at once. The answer was simple: She could put two separate purchases of 25 on her credit card (below the limit), but if she went to 50, it had to go through purchasing, which would take 4 to 6 weeks.

Could I have forced her to pay the extra $10 per copy to avoid going through purchasing? Sure, because it would save her more than $10 in time. Does it cost me twice as much to fill two orders as it does to do one? You bet. Did I give her the deal anyway? I had to because I despise idiotic bureaucracy.

How Much Are Delays in Purchasing Costing Your Company?

Another state agency wanted to buy an enterprise license that would save it a significant amount over any of my other discounts. All the agency had to do was to get purchasing to issue the order. A purchasing agent called me and asked if I sell my product through resellers. I do, but not my enterprise licenses. He didn't care. He then had to call three resellers and see if he could get a better price through them. So each of the three resellers had to call me to find out about pricing. Of course, they asked about a quantity half the size of the license. So I had to give them a heads-up that purchasing was screwing around with all of us. Most of them admitted that the state's purchasing department had given them fits in the past.

Of course, then the purchasing agent fed these quotes back to the buyer without mentioning that there was a reduction in quantities. So the buyer had to call me, thinking I was trying to finagle something.

How will this all turn out? I don't know because it all started two weeks ago when the customer tried to order 100 copies and I tried to do the customer a favor by upgrading to the lowest enterprise license, which would save the customer money. Meanwhile, the people who need the software aren't able to do anything. What did all this churn cost the customer? More than the license is worth, I'm sure.

Is a foolish constraint in one department such as purchasing driving the rest of your company to drink? Is it driving your suppliers to consider firing you as a customer because you cost too much to manage? Is micromanagement in one department killing your ability to perform in the marketplace?

Realign your measures and goals. Purchasing, for example, should be rewarded on speed to issue a purchase order and minimizing total cost. If purchasing only counts the pennies saved on each order and not the total cost to the organization of the delays involved, then it will optimize the goal set for itself. This is true of any department.

What are the idiotic goals your department lives by that constrain the overall productivity of the company? Does IT take too many months to implement a software change? Does billing take too long to issue an invoice? Is accounts payable taking too long to issue payments, causing collection calls by suppliers and delays in new shipments?

You're not a silo any more. Get over it. *Align the goals throughout the company to maximize your speed, quality, and profitability.* Eliminate the ones that slow you down or hold you back.

HONOR YOUR PROGRESS

Clients tell me that it's often hard to sustain the momentum of Lean Six Sigma. When I ask what they've done to recognize and reward teams for success, they often hesitate and then mumble something about money.

We know that you get more of what you reward, and we know from human resources studies that lack of money is a demotivator, not a motivator. So what can we do to reinforce Lean Six Sigma behaviors?

To answer this question, I'd like you to think back to the times when you felt most recognized for your contributions. What did your leaders say or do that let you know that they fully understood your accomplishments?

One thing I've noticed, monetary rewards are soon spent and quickly forgotten, but something tangible often remains long afterward to remind employees of their contribution. My last year at the phone company, I worked on a project that helped to save millions of dollars. All the team members and I were treated to an off-site retreat, and we were each given a leather jacket from Warner Brothers that had all the Looney Tunes characters on the back. And our team members were every bit as diverse as those characters, but we'd found a way to work together to achieve outrageous results. I've also seen teams rewarded with popcorn or pizza parties, votive candles, and just plain time with the executive team to present their story.

I think the key is to find a unique way to recognize the team that reflects the team's sense of values and its contribution to the success of the business. It's like picking out a gift for a friend or family member. You don't want the same old thing everybody else has; you want something special that the person will remember.

What are you doing to recognize and reinforce the spread of Lean Six Sigma in your organization?

THE HARD WORK IS SOFT

While figuring out what to fix can be a "slog," in the *The Toyota Way* (McGraw-Hill, 2004), author Jeffery Liker admits that most of the progress occurs through detailed, painstaking problem solving (i.e., a slog). The biggest challenge is "getting employees to accept that how they've always done things may not be the best way." Another soft challenge is tearing down the walls between divisions to implement some of the changes.

As with any change, the hardest part is getting the people involved to agree to the change. And the best way to do that is to involve them in the analysis of the problem and creation of the solution because then they own the change.

How can Lean Six Sigma boost your profits?

LEAN SIX SIGMA ROLES

I have found that for any improvement team to succeed, the team members need three things:

1. *A sponsor in management*
2. *A higher-level-management sponsor (or champion)*
3. *A change agent or facilitator*

Traditional Lean Six Sigma Roles

Champions actively sponsor and provide leadership for Lean Six Sigma projects. *Master black belts* (MBBs) oversee the Lean Six Sigma projects. If your company is big enough to have more than 10 improvement projects running at one time, you probably need a master black belt. *Black belts* (BBs) facilitate, lead, and coach improvement teams *full time*. The American Society for Quality (ASQ) has a body of knowledge (BOK) for black belts at www.asq.org/certification/six-sigma/bok.html. *Green belts* (GBs) work on improvement projects part time; go to www.asq.org/certification/six-sigma-green-belt/bok.html. *Process owners* manage cross-functional, mission-critical business processes. They have the responsibility and authority to change the process.

GET A FASTER, BETTER, CHEAPER HOSPITAL

Money belts laser-focus improvements and drive bottom-line results. I'd like you to consider that starting with the whole Lean Six Sigma hierarchy is unnecessary when you start. It's something you grow as Lean Six Sigma takes root and grows.

Using SWAT teams laser-focused on mission-critical improvements, it's possible to get dramatically faster, better, and cheaper in five days or less. It doesn't have to take forever, but it does require the right people focused on a well-defined problem.

Would you rather spend five days in training or five days making dramatic progress on specific business issues? And learn the essential methods and tools of Lean Six Sigma as a by-product of improvement?

As you can see, there are lots of ways to fail at integrating Lean Six Sigma into your culture. There are many ways for Lean Six Sigma to kill your productivity and profits and even your company if you go overboard on implementation.

Instead, pilot a few projects. Establish a track record of success, and expand into increasingly important improvement projects.

Where to Implement Lean Six Sigma in Hospitals

- Admissions/discharge
- Bed management
- Emergency department
- Operating room
- Pharmacy

- Radiology
- Nursing units
- Kitchen
- Purchasing/supply
- Information systems
- Administration

Bibliography

Automotive Industry Action Group (AIAG). *Statistical Process Control—SPC*, 2nd ed. Detroit: AIAG, 2005.

AIAG. *Measurement Systems Analysis—MSA*, 3rd ed. Detroit: AIAG, 2005.

Adrian, Nicole. "A Gold Medal Solution," *Quality Progress*, March 2008, pp. 45–50.

Arthur, Jay. *How to Motivate Everyone*, Denver: LifeStar, 2002.

Arthur, Jay. *Six Sigma Simplified*. Denver: LifeStar, 2004.

Arthur, Jay. *Lean Six Sigma Demystified*. New York: McGraw-Hill, 2007.

Artis, Spencer E. "Six Sigma Kick-Starts Starwood," *BusinessWeek Online*, August 31, 2007.

Auge, Karen. "Thinking Like Factory Helps Heal Hospital's Bottom Line," *Denver Post*, July 28, 2010.

Bala, S. "Lean Triage for Hospital ERs," *Quality Digest*, March 2009, pp. 22–24.

Barker, Joel. *Future Edge*. New York: William Morrow, 1992.

Bates, David W., et al. "Medication Safety Technologies: What Is and Is Not Working," *Patient Safety & Quality Healthcare*, July–August 2009, pp. 22–27.

Berry, Leonard Eugene. *Management Accounting Demystified*. New York: McGraw-Hill, 2006.

Bohmer, Richard M. J. "Fixing Health Care on the Front Lines," *HBR*, April 2010, pp. 63–69.

Bossidy, Larry, and Ram Charan. *Execution: The Discipline of Getting Things Done*, New York: Crown Business, 2002.

Buckingham, Marcus. *The One Thing You Need to Know*. New York: Free Press, 2005.

Chaudhry, Imarn. "Surgical Infection Prevention," *iSixSigma Magazine*, September–October 2008, pp. 49–51.

Christiansen, Clayton. *The Innovator's Dilemma*. Boston: Harvard Business School Press, 2000.

Christiansen, Clayton. *The Innovator's Prescription*. New York: McGraw-Hill, 2009.

Cialdini, Robert B. *Influence: The Psychology of Persuasion*. New York: William Morrow, 1993.

Clance, Carolyn M. "One Decade after To Err Is Human," *Patient Safety & Quality Healthcare*, September–October 2009, pp. 8–10.

Clarke, John R., et al. "Getting Surgery Right," *Annals of Surgery*; available at http://cme.medscape.com/viewarticle/565349.

Collins, Jim. *Good to Great: Why Some Companies Make the Leap... and Others Don't*. New York: HarperCollins, 2001.

Collins, Jim and Jerry I. Porras. *Built to Last: Successful Habits of Visionary Companies*. New York: HarperCollins, 1997.

Colvin, Geoff. *Talent Is Overrated*. New York: Portfolio, 2008.

Cyr, Jay, et al. "Sustaining and Spreading Reduced Door-to-Balloon Times for ST-Segment Elevation Myocardial Infarction Patients," *The Joint Commission Journal on Quality and Patient Safety*, June 2009, pp. 297–306.

Downes, Larry, and Chunka Mui. *Unleashing the Killer App*. Boston: Harvard Business School Press, 1998.

Dusharme, Dirk. "Six Sigma Survey," *Quality Digest*, February 2003 and September 2004.

Esimai, Grace. "Lean Six Sigma Reduces Medication Errors," *Quality Progress*, April 2005, pp. 51–57.

Farzad, Roben. "The Toyota Enigma," *BusinessWeek*, July 10, 2006, p. 30.

Furman, Cathie, and Robert Caplan. "Applying the Toyota Production System: Using a Patient Safety Alert System to Reduce Error," *The Joint Commission Journal on Quality and Patient Safety* 33(7):376–386, 2007.

Gawande, Atul. *The Checklist Manifesto*. New York: Henry Holt, 2009.

Gawande, Atul. *Better*. New York: Henry Holt, 2007.

Gawande, Atul. *Complications*. New York: Henry Holt, 2003.

George, Michael, James Works, and Kimberly Watson-Hemphill. *Fast Innovation*. New York: McGraw-Hill, 2005.

Gilbert, Lindsey, et al. "Aligning Hospital and Physician Performance Incentives: A Shared Success Model," *The Joint Commission Journal on Quality and Patient Safety* 34(12):703–706, 2008.

Gladwell, Malcolm. *The Tipping Point*. Boston: Little Brown, 2002.

Gladwell, Malcolm. *Outliers: The Story of Success*, New York: Little, Brown and Company, 2008.

Godin, Seth. *Unleashing the Ideavirus*. New York: Hyperion, 2001.

Goldratt, Eliyahu M. and Jeff Cox. *The Goal: A Process of Ongoing Improvement*. Great Barrington, MA: North River Press, 1984.

Grout, John. *Mistake-Proofing the Design of Health Care Processes*. Rockville, MD: Agency for Healthcare Research and Quality (AHRQ) 2007.

Hall, Kenji. "No One Does Lean Like the Japanese," *BusinessWeek*, July, 10, 2006, pp. 40–41.

Harrington, H. James. "Just the Facts on Health Care," *Quality Digest*, March 2006, p. 14.

Harrington, H. James. "How Serious Is the Health Care Problem?" *Quality Digest*, April 2006, p. 14.

"Health Care Needs a New Kind of Hero," *HBR*, April 2010, pp. 60–61.

Health Grades, "The Fifth Annual Health Grades Patient Safety in American Hospitals Study," *Health Grades*, Golden, 2008.

Heath, Dan, and Chip Heath. *Made to Stick*. New York: Random House, 2007.

Heath, Dan, and Chip Heath. *Switch*. New York: Random House, 2010.

Horst, Mathilda, et al. "A Tight Glycemic Control Initiative in a Surgical Intensive Care Unit and Hospital Wide," *The Joint Commission Journal on Quality and Patient Safety* 36(7):291–300, 2010.

"Hospitals See Benefits of Lean and Six Sigma," American Society for Quality (ASQ) Benchmarking Study, 2009; available at www.asq.org/media-room/press-releases/2009/20090318-hospitals-see-benefits-lss.html.

Institute of Medicine. *To Err Is Human*. Washington, DC: National Academy Press, 2000.

Jencks, Stephen F., et al. "Rehospitalizations Among Patients in the Medicare Fee-for-Service Program," *New England Journal of Medicine* 360:1418–1428, 2009.

Jones, Dell. "Hospital CEOs Find Ways to Save," *USA Today*, September 10, 2009.

Kaplan, Robert S., and David P. Kaplan. *The Balanced Scorecard*. Boston: Harvard Business School Press, 1996.

Kaplan, Robert S., and David P. Kaplan. *The Strategy Focused Organization*. Boston: Harvard Business School Press, 2001.

Kauffman, Stuart. *At Home in the Universe*. New York: Oxford University Press, 1995.

Kay, Tan. "Room for Improvement," *Six Sigma Forum Magazine*, November 2009.

Keen, Peter. *The Process Edge: Creating Value Where It Counts*. Boston: Harvard Business School Press, 1997.

Kim, Christopher S., et al. "Implementation of Lean Thinking: One Health System's Journey," *The Joint Commission Journal on Quality and Patient Safety*, August 2009, pp. 405–413.

Kohn, Linda T., Janet M. Corrigan, and Molla S. Donaldson, eds. *To Err Is Human*. Washington, DC: National Academy Press, 2000.

Krasner, Jeffrey. "New Medicine for What Ails Hospitals," *Boston Globe*, January 28, 2008.

Krzykowski, Brett. "In a Perfect World," *Quality Progress*, June 2009, pp. 33–39.

Lashinsky, Adam. "The Genius Behind Steve," *Fortune*, November 10, 2008; available at http://money.cnn.com/2008/11/09/technology/cook_apple.fortune/index.htm.

Lee, Thomas H. "Turning Doctors into Leaders," *Harvard Business Review*, April 2010, pp. 51–58.

Levitt, Steven D. and Stephen J. Dubner. *Freakonomics: A Rogue Economist Explores the Hidden Side of Everything*. New York: HarperCollins, 2005.

Lewis, Michael. *Moneyball*. New York: Norton, 2003.

Liker, Jeffrey. *The Toyota Way*. New York: McGraw-Hill, 2004.

Luntz, Dr. Frank. *Words that Work: It's Not What You Say, It's What People Hear*. Hyperion, 2007.

Mauboussin, Michael J. *Think Twice*. Boston: Harvard Business School Press, 2009.

Meyer, Christopher. *Relentless Growth—How Silicon Valley Innovation Strategies Can Work in Your Business*. New York: Free Press, 1998.

Miller, Ken. *We Don't Make Widgets—Overcoming the Myths That Keep Government from Radically Improving*. New York: Governing Books, 2006.

Moore, Geoffrey. *Crossing the Chasm*. New York: Harper Business, 1999.

Ohno, Taiichi. *Toyota Production System*. New York: Productivity Press, 1988.

Pelczarski, Kathryn, and Cynthia Wallace. "Hospitals Collaborate to Prevent Falls," *Patient Safety & Quality Healthcare*, November–December 2008, pp. 30–36.

Powers, Donna, and Mary Paul. "Healthcare Department Reduces Cycle Time and Errors," *Six Sigma Forum Magazine*, February 2008, pp. 30–34.

Rashidee, Ali, et al. "High-Alert Medications: Error Prevalence and Severity," *Patient Safety & Quality Healthcare*, July–August 2009, pp. 16–19.

Reibling, Nancy B., et al. "CT Scan Throughput," *iSixSigma Magazine*, January–February 2010, pp. 49–54.

Reibling, Nancy B., et al. "Toward Error Free Lab Work," *Six Sigma Forum Magazine*, November 2004, pp. 23–29.

Rogers, Everett. *Diffusion of Innovations*, 4th ed. New York: Free Press, 1995.

Ronen, Boaz, et al. *Focused Operations Management for Health Services Organizations*. San Francisco: Jossey-Bass, 2006.

Schank, Roger. *Lessons in Learning, e-Learning and Training*. New York: John Wiley & Sons, 2005.

Schmidt, Elaine. "Crystal Clear," *iSixSigma Magazine*, March–April 2008.

Schmidt, Elaine. "RX for Success," *iSixSigma Magazine*, May–June 2008.

Schmidt, Elaine. "From the Bottom Up," *iSixSigma Magazine*, September–October 2008, pp. 24–32.

Scott, R. Douglas, II. "Direct Medical Costs of Healthcare-Associated Infections in U.S. Hospitals and the Benefits of Prevention, National Center for Preparedness," *Detection and Control of Infectious Diseases*, March 2009.

Shannon, Richard P., et al. "Using Real-Time Problem Solving to Eliminate Central Line Infections," *The Joint Commission Journal on Quality and Patient Safety* 32(9):479–487, 2006.

Shenkar, Oded. "Imitation Is More Valuable than Innovation," *HBR*, April 2010, pp. 28–29.

Stauk, George, and Thomas M. Hout. *Competing Against Time*. New York: Free Press, 1990

Stock, Greg. "Taking Performance to a Higher Level," *Six Sigma Forum Magazine*, May 2002, pp. 23–26.

Tufte, Edward. *Visual Explanations*. Cheshire, CT: Graphic Press, 1997.

Tufte, Edward. *Envisioning Information*. Cheshire, CT: Graphic Press, 1990.

Tukey, J. W. "A Quick, Compact, Two-Sample Test to Duckworth's Specifications," *Technometrics* 1(1), 1959.

Weed, Julie. "Factory Efficiency Comes to the Hospital," *New York Times*, July 9, 2010.

Welch, Jack and Suzy Welch. *Winning: The Answers: Confronting 74 of the Toughest Questions in Business Today*. New York: HarperCollins, 2006.

Wennecke, Gette. "Kaizen—Lean in a Week," August 2008; available at www.mlo-online.com.

Widner, Tracy, and Mitch Gallant. "A Launch to Quality," *Quality Progress*, February 2008, pp. 38–43.

Winston, Stephanie. *The Organized Executive*. New York: Schuster, 1996.

Womack, James P., and Daniel T. Jones. *Lean Thinking*. New York: Simon & Schuster, 1996.

Index